# SAI SPEED MATH ACADEMY

# ABACUS MIND MATH

## Excel at Mind Math with Soroban, a Japanese Abacus

# LEVEL – 1

## WORKBOOK 2 OF 2

**PUBLISHED BY SAI SPEED MATH ACADEMY**

USA

www.abacus-math.com

Published in the United States of America by SAI Speed Math Academy, 2014

The Library of Congress has cataloged this book under this catalog number:
Library of Congress Control Number:  2014907005


ISBN of this edition: 978-1-941589-02-1

Thanks to **Abiraaman Amarnath** for his valuable contribution towards the development of this book.

Edited by: WordPlay
www.wordplay.com
Front Cover Image: © [Yael Weiss] / Dollar Photo Club
www.abacus-math.com
Printed in the United States of America

Our Heartfelt Thanks to:

Our

Higher Self,

Family,

Teachers,

And Friends

for the support, guidance and confidence they gave us to…

…become one of the rare people who don't know how to quit. (–Robin Sharma)

# KIND REQUEST

*We believe knowledge is sacred.*

*We believe that knowledge has to be shared.*

We could have monopolized our knowledge by franchising our work and creating wealth for ourselves. However, we choose to publish books so we can reach more parents and teachers who are interested in empowering their children with mind math at a very affordable cost and with the convenience of teaching at home.

Please help us know that we made the right decision by publishing books.

- ❖ We request that you please buy our books first hand to motivate us and show us your support.
- ❖ Please do not buy used books.
- ❖ We kindly ask you to refrain from copying this book in any form.
- ❖ Help us by introducing our books to your family and friends.

We are very grateful and truly believe that we are all connected through these books. We are very grateful to all the parents who have called in or emailed us to show their appreciation and support.

Thank you for trusting us and supporting our work.

*With Best Regards,*

*SAI Speed Math Academy*

Dear Parents and Teachers,

Thank you very much for buying this workbook. We are honored that you choose to use this workbook to help your child learn math and mind math using the Japanese abacus called the "Soroban". This is our effort to bring a much needed practice workbook to soroban enthusiasts around the world.

This book is the product of over six years of intense practice, research, and analysis of soroban. It has been perfected through learning, applying, and teaching the techniques to many students who have progressed and completed all six levels of our course successfully.

We are extremely grateful to all who have been involved in this extensive process and with the development of this book.

We know that with *effort*, *commitment* and *tenacity*, everyone can learn to work on soroban and succeed in mind math.

We wish all of you an enriching experience in learning to work on abacus and enjoying mind math excellence!

We are still learning and enjoying every minute of it!

## GOAL AFTER COMPLETION OF LEVEL 1 – WORKBOOKS 1 AND 2

On successful completion of the two workbooks students would be able to:

1. Add any two digit numbers that does not involve carry-over or regrouping problems.
2. Subtract any two digit numbers that does not involve borrowing or regrouping problems.

## HELPFUL SKILLS

- Know to read and write numbers 0-99
- Know to identify place value of numbers
- Must have completed **LEVEL – 1 WORKBOOK 1 OF 2**.

## INSTRUCTION BOOK FOR PARENTS/TEACHERS

Workbooks do not contain any instruction on how to work with an abacus. All instructions are in the Instruction Book for parents which is **sold separately** under the title:

**Abacus Mind Math Level – 1 Instruction Book – ISBN: 978-1-941589-00-7**

# HIGHER LEVEL

Thank you very much for using our LEVEL – 1 books. After you complete LEVEL – 1 by completing Workbook 1 of 2 and Workbook 2 of 2 please proceed to use our LEVEL – 2 books to continue your abacus training.

LEVEL – 2 books are **sold separately** under the title:

**Abacus Mind Math Level – 2 Instruction Book – ISBN: 978-1-941589-03-8**

**Abacus Mind Math Level – 2 Workbook 1 of 2 – ISBN: 978-1-941589-04-5**

**Abacus Mind Math Level – 2 Workbook 2 of 2 – ISBN: 978-1-941589-05-2**

## WE WOULD LIKE TO HEAR FROM YOU!

Please visit our Facebook page at https://www.facebook.com/AbacusMindMath. Contact us through http://www.abacus-math.com/contactus.php or email us at **info@abacus-math.com**.

## We Will Award Your Child a Certificate Upon Course Completion:

Once your child completes the test given at the back of the workbook – 2, please upload pictures of your child with completed test and marks scored on our Facebook page at https://www.facebook.com/AbacusMindMath, and at our email address: http://www.abacus-math.com/contactus.php

Provide us your email and we will email you a personalized certificate for your child. Please include your child's name as you would like for it to appear on the certificate.

## LEARNING INSTITUTIONS AND HOME SCHOOLS

If you are from any public, charter or private school, and want to provide the opportunity of learning mind math using soroban to your students, please contact us. This book is a good teaching/learning aid for small groups or for one on one class. Books for larger classrooms are set up as 'Class work books' and 'Homework books'. These books will make the teaching and learning process a smooth, successful and empowering experience for teachers and students. We can work with you to provide the best learning experience for your students.

If you are from a home school group, please contact us if you need any help.

# Contents

SAI Speed Math Academy

# HOW TO USE THIS WORKBOOK

Use this workbook after completing ABACUS MIND MATH LEVEL 1 - WORKBOOK 1 of 2. Work in the order given.

Each and every child is unique in his/her ability to learn. Sometimes a lesson might have to be repeated to get better understanding. You can erase the answers and redo the same lesson. On the other hand you may choose to move through the lessons quicker if child is an easy learner.

Each week's work is grouped together. Finish all the pages under each week before moving on to the next week's work. Work is divided for 5 days and you may choose to combine any number of days if you want to finish in shorter time. But, be very careful when you choose to do this, because children get overwhelmed easily when introduced to too many new concepts in a short time without having enough time to understand and practice. Use your best judgment since you know your child's temperament and learning capabilities.

After finishing a DAY's work check the answer and redo the problems with the wrong answer.

Wish you all the best!

**FINGERING** – Correct fingering is very important so, practice moving earth beads and heaven beads using the correct fingers.

**WEEK 11** – Dictation introduced

**WEEKS 11 to 14** – Small friend's formulas for +2, +3, -2, and -3 are introduced. First complete the rows that are to be worked with abacus. After this, move on to the section where it says to work in mind and try working on them in mind. Finish the day's work with dictation.

**WEEKS 15** – Skill building week – no new formula is introduced during a skill building week. Concepts learnt thus far are practiced.

**WEEKS 16 to 17** – Continue learning with small friends formula of +4 and -4 formulas.

**WEEK 18** – Skill building week

**WEEK 19** – Using of small friends formula on hundred's place value rod introduced

**WEEK 20** – Skill building week

**END OF LEVEL TEST** – Wish you all the best!

## KEEPING TRACK

**TIME:** Make note of the time it takes your student to finish each day's work.

**GRADES:** Correct their work and calculate grade. Let your student color the stars next to each day's work. This will keep them engaged and encouraged.

**GOAL:** As student progresses through the week they should be able to do their work in less time with more accuracy.

## HOW TO CALCULATE GRADE?

$$\frac{\text{Number of correct problems}}{\text{Total number of problems}} \times 100 = \text{Percentage scored}$$

## GRADES

| GRADE | PERCENTAGE | | STAR COLOR |
|-------|-----------|--|-----------|
| A+ | 96-100 | EXTRAORDINARY | GOLD |
| A | 91-95 | EXCELLENT | |
| B+ | 86-90 | AWESOME | SILVER |
| B | 81-85 | GOOD | |
| C | 76-80 | CONGRATULATIONS | BROWN |
| C+ | 70-75 | CONGRATULATIONS | |

# FINGERING

## JOB OF THE THUMB (1):

A. Used to push the Earth Beads up to the beam, adding them to the game (ADD).

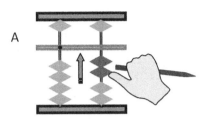

## JOB OF THE POINTER FINGER (3):

1. Used to push the Earth Beads away from the beam, removing them from the game (MINUS).
2. Used to push the Heaven bead down to touch the beam, adding it to the game (ADD).
3. Used to push the Heaven bead away from the beam, removing it from the game (MINUS).

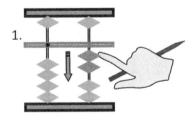 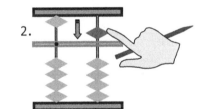 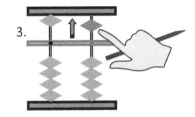

## JOB OF THE OTHER THREE FINGERS:

Use your other three fingers to hold your pencil with the point facing down or away from you.

# WEEK 11 – LESSON 11 – INTRODUCING +2 CONCEPT AND DICTATION

## LESSON 11 – PRACTICE WORK

$$+2 = + 5 - 3$$

**DAY 1 – MONDAY**　　Accuracy _____ /32 ☆　　Use Abacus

| 1 | 2 | 3 | 4 | 5 | 6 | 7 | 8 | |
|---|---|---|---|---|---|---|---|---|
| 33 | 30 | 24 | 23 | 33 | 13 | 22 | 30 | |
| 02 | 20 | 22 | 12 | 22 | 12 | 22 | 20 | |
| - 05 | 04 | 20 | 11 | - 55 | 23 | 12 | 44 | 11:1 |
| 54 | 02 | - 15 | 23 | 44 | 20 | - 55 | 01 | |
| | | | | | | | | |

| 1 | 2 | 3 | 4 | 5 | 6 | 7 | 8 | |
|---|---|---|---|---|---|---|---|---|
| 21 | 30 | 44 | 44 | 84 | 64 | 33 | 34 | |
| 12 | 20 | 11 | 12 | 12 | 22 | 12 | 21 | |
| 21 | 24 | 22 | 20 | - 55 | - 55 | 21 | - 15 | 11:2 |
| 12 | 01 | - 66 | 03 | 26 | 13 | - 60 | 29 | |
| | | | | | | | | |

| 1 | 2 | 3 | 4 | 5 | 6 | 7 | 8 | |
|---|---|---|---|---|---|---|---|---|
| 55 | 34 | 44 | 20 | 43 | 44 | 74 | 14 | |
| - 11 | 21 | 21 | 24 | 22 | 21 | 02 | 21 | |
| 21 | 31 | 01 | - 30 | - 10 | 13 | - 25 | 22 | 11:3 |
| - 11 | 13 | - 50 | 01 | - 15 | - 57 | - 11 | - 16 | |
| | | | | | | | | |

| 1 | 2 | 3 | 4 | 5 | 6 | 7 | 8 | |
|---|---|---|---|---|---|---|---|---|
| 35 | 34 | 34 | 43 | 86 | 37 | 48 | 59 | |
| - 21 | 21 | 12 | 22 | - 21 | 22 | 20 | - 17 | |
| 72 | - 15 | 21 | - 11 | - 11 | - 56 | - 55 | 22 | 11:4 |
| - 35 | - 30 | - 65 | 02 | 42 | 02 | 02 | 32 | |
| | | | | | | | | |

www.abacus-math.com

## DAY 2 – TUESDAY

| 1 | 2 | 3 | 4 | 5 | 6 | 7 | 8 |
|---|---|---|---|---|---|---|---|
| 34 | 23 | 38 | 42 | 48 | 60 | 54 | 34 |
| 22 | 22 | 21 | 22 | 20 | 20 | 11 | 22 |
| 11 | 11 | - 55 | - 51 | - 55 | 04 | 30 | - 10 |
| 31 | 20 | 91 | 02 | 72 | - 72 | - 01 | 20 |
|  |  |  |  |  |  |  |  |

11:5

| 1 | 2 | 3 | 4 | 5 | 6 | 7 | 8 |
|---|---|---|---|---|---|---|---|
| 43 | 44 | 33 | 32 | 40 | 54 | 45 | 65 |
| - 13 | 22 | 02 | 25 | 03 | 02 | 24 | - 11 |
| 27 | 22 | 20 | - 06 | 25 | - 10 | - 54 | 32 |
| - 05 | 11 | - 11 | 33 | - 51 | 21 | - 01 | - 51 |
|  |  |  |  |  |  |  |  |

11:6

## DAY 3 – WEDNESDAY

| 1 | 2 | 3 | 4 | 5 | 6 | 7 | 8 |
|---|---|---|---|---|---|---|---|
| 63 | 44 | 53 | 44 | 53 | 33 | 34 | 34 |
| 12 | 22 | 22 | 22 | 12 | 22 | 21 | 12 |
| 12 | 21 | - 51 | - 11 | 21 | - 11 | 04 | 20 |
| - 57 | - 32 | - 24 | - 11 | - 35 | 22 | - 18 | - 66 |
|  |  |  |  |  |  |  |  |

11:7

| 1 | 2 | 3 | 4 | 5 | 6 | 7 | 8 |
|---|---|---|---|---|---|---|---|
| 24 | 44 | 23 | 13 | 34 | 21 | 37 | 24 |
| 12 | 22 | 12 | 31 | 21 | 23 | 21 | 52 |
| 12 | - 55 | 23 | 55 | 44 | 21 | 20 | - 21 |
| 20 | 77 | - 11 | - 78 | - 91 | - 51 | - 78 | - 51 |
|  |  |  |  |  |  |  |  |

11:8

## DAY 4 – THURSDAY

TIME: _____min _____sec    Accuracy _____/16    ☆

| 1 | 2 | 3 | 4 | 5 | 6 | 7 | 8 |
|---|---|---|---|---|---|---|---|
| 54 | 43 | 14 | 37 | 59 | 43 | 11 | 47 |
| 01 | 22 | 22 | 22 | - 16 | 12 | 22 | 22 |
| - 11 | - 11 | - 21 | - 16 | 52 | 12 | 21 | - 55 |
| 22 | 21 | - 11 | 12 | - 41 | - 55 | - 54 | 02 |
|  |  |  |  |  |  |  |  |

11:9

| 1 | 2 | 3 | 4 | 5 | 6 | 7 | 8 |
|---|---|---|---|---|---|---|---|
| 24 | 35 | 43 | 23 | 44 | 27 | 34 | 48 |
| 22 | 21 | 21 | 22 | 20 | 22 | 21 | 21 |
| - 35 | 12 | 11 | - 31 | - 13 | 10 | - 11 | - 16 |
| 87 | - 67 | - 55 | 72 | - 10 | - 19 | 52 | - 13 |
|  |  |  |  |  |  |  |  |

11:10

## DAY 5 – FRIDAY

TIME: _____min _____sec    Accuracy _____/16    ☆

| 1 | 2 | 3 | 4 | 5 | 6 | 7 | 8 |
|---|---|---|---|---|---|---|---|
| 54 | 23 | 37 | 43 | 55 | 36 | 44 | 34 |
| 22 | 12 | 22 | 12 | - 11 | 22 | 25 | 21 |
| 20 | 12 | - 15 | - 11 | 22 | - 13 | - 55 | 23 |
| - 15 | 22 | - 44 | 22 | - 55 | 22 | 72 | - 76 |
|  |  |  |  |  |  |  |  |

11:11

| 1 | 2 | 3 | 4 | 5 | 6 | 7 | 8 |
|---|---|---|---|---|---|---|---|
| 93 | 47 | 89 | 64 | 32 | 62 | 72 | 21 |
| - 40 | 12 | - 35 | 25 | 22 | 12 | 15 | 22 |
| 32 | - 15 | - 11 | - 54 | - 14 | 11 | - 52 | 22 |
| - 51 | 22 | 22 | 20 | - 40 | - 35 | 24 | 33 |
|  |  |  |  |  |  |  |  |

11:12

# LESSON 11 – MIND MATH PRACTICE

 Visualize

## DAY 1 – MONDAY

Accuracy _____/10

| 1 | 2 | 3 | 4 | 5 | 6 | 7 | 8 | 9 | 10 |
|---|---|---|---|---|---|---|---|---|---|
| 04 | 30 | 31 | 45 | 14 | 34 | 04 | 30 | 54 | 44 |
| 02 | 20 | 23 | 20 | 12 | 20 | 02 | 20 | 02 | 12 |
| - 06 | 07 | - 04 | - 01 | - 01 | 11 | - 01 | - 10 | - 10 | 22 |
|  |  |  |  |  |  |  |  |  |  |

11:13

## DAY 2 – TUESDAY

Accuracy _____/10

| 1 | 2 | 3 | 4 | 5 | 6 | 7 | 8 | 9 | 10 |
|---|---|---|---|---|---|---|---|---|---|
|  |  |  |  |  | 25 | 33 | 24 | 34 | 46 |
| 34 | 04 | 25 | 43 | 23 | - 11 | 12 | 20 | 12 | 10 |
| 12 | 22 | - 01 | 12 | 22 | 02 | - 01 | 20 | - 11 | - 05 |
| - 11 | 20 | 02 | - 01 | 10 | - 05 | - 20 | - 13 | - 01 | - 10 |
|  |  |  |  |  |  |  |  |  |  |

11:14

## DAY 3 – WEDNESDAY

Accuracy _____/10

| 1 | 2 | 3 | 4 | 5 | 6 | 7 | 8 | 9 | 10 |
|---|---|---|---|---|---|---|---|---|---|
|  |  |  |  |  | 40 | 44 | 24 | 53 | 44 |
| 13 | 12 | 44 | 15 | 45 | 20 | 20 | 22 | - 03 | 22 |
| 11 | 32 | 01 | - 01 | - 01 | 24 | 30 | 21 | 09 | 22 |
| 12 | 02 | 20 | 12 | 20 | - 72 | - 22 | 21 | - 15 | - 85 |
|  |  |  |  |  |  |  |  |  |  |

11:15

## DAY 4 – THURSDAY

Accuracy _____/10

| 1 | 2 | 3 | 4 | 5 | 6 | 7 | 8 | 9 | 10 |
|---|---|---|---|---|---|---|---|---|---|
| 43 | 85 | 33 | 44 | 54 | 42 | 33 | 34 | 44 | 46 |
| 02 | - 30 | 02 | 02 | 02 | 12 | 22 | 20 | 15 | 20 |
| 20 | - 01 | - 11 | - 21 | - 10 | 12 | - 01 | 01 | - 11 | - 11 |
| - 01 | - 10 | - 02 | - 21 | - 06 | - 55 | 40 | - 10 | - 15 | - 11 |
|  |  |  |  |  |  |  |  |  |  |

11:16

Accuracy _____/20 ☆

| 1 | 2 | 3 | 4 | 5 | 6 | 7 | 8 | 9 | 10 |
|---|---|---|---|---|---|---|---|---|---|
| 04 | 30 | 40 | 04 | 35 | 56 | 94 | 21 | 55 | 57 |
| 02 | 25 | 03 | 02 | 24 | - 11 | - 30 | 23 | - 11 | - 10 |
| 40 | - 05 | 02 | 30 | - 10 | 20 | 02 | 05 | 22 | 21 |
| 20 | 01 | 20 | 20 | - 40 | - 60 | - 15 | 20 | - 51 | - 16 |
| | | | | | | | | | |

11:17 © SAI Speed Math Academy, USA

| 1 | 2 | 3 | 4 | 5 | 6 | 7 | 8 | 9 | 10 |
|---|---|---|---|---|---|---|---|---|---|
| 19 | 55 | 44 | 54 | 35 | 36 | 54 | 51 | 75 | 65 |
| - 05 | - 10 | 02 | 12 | - 21 | - 21 | - 10 | 03 | - 71 | 10 |
| 22 | 11 | - 05 | - 10 | 01 | 70 | 02 | - 10 | 02 | - 51 |
| 10 | - 50 | 20 | - 05 | 20 | - 01 | - 30 | 02 | - 05 | - 01 |
| | | | | | | | | | |

11:18 © SAI Speed Math Academy, USA

## SHAPE PUZZLE

Find out what each shape is worth.

Suggestion: Take as many numbers of beans, blocks or coins as the right side of the equation shows and share them on the shapes to the left side of the equation.

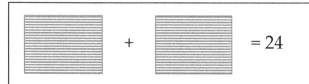

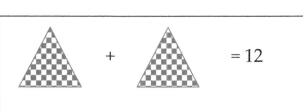

www.abacus-math.com

## LESSON 11 – DICTATION

**DICTATION:** Dictation is when teacher or parent calls out a series of numbers and the child listens to the numbers and does the calculation in mind or on the abacus.

**DO 6 PROBLEMS A DAY** and write answers below.

Students: Use abacus to calculate dictation problems this week.

Teachers: Dictate problems from mind math part of this week's homework.

| 1 | 2 | 3 | 4 | 5 | 6 | 7 | 8 | 9 | 10 |
|---|---|---|---|---|---|---|---|---|----|
|   |   |   |   |   |   |   |   |   |    |

| 11 | 12 | 13 | 14 | 15 | 16 | 17 | 18 | 19 | 20 |
|----|----|----|----|----|----|----|----|----|----|
|    |    |    |    |    |    |    |    |    |    |

| 21 | 22 | 23 | 24 | 25 | 26 | 27 | 28 | 29 | 30 |
|----|----|----|----|----|----|----|----|----|----|
|    |    |    |    |    |    |    |    |    |    |

## WEEK 11 – SKILL BUILDING

Visualize the numbers on the beam in your mind and draw it to represent the numbers given.

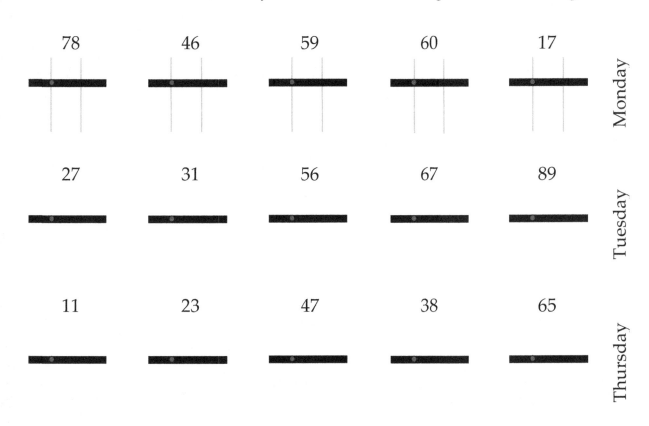

Monday: 78   46   59   60   17

Tuesday: 27   31   56   67   89

Thursday: 11   23   47   38   65

# WEEK 12 – LESSON 12 – INTRODUCING –2 CONCEPT

## LESSON 12 – PRACTICE WORK

$$-2 = -5 + 3$$

Accuracy _____ /32 ☆

Use Abacus

### DAY 1 – MONDAY

| 1 | 2 | 3 | 4 | 5 | 6 | 7 | 8 | |
|---|---|---|---|---|---|---|---|---|
| 04 | 54 | 55 | 50 | 35 | 60 | 50 | 55 | 12:1 |
| 01 | - 23 | 10 | - 20 | 12 | 05 | 15 | - 22 | |
| - 02 | - 11 | - 20 | 10 | - 21 | 04 | 02 | 55 | |
| 15 | 26 | - 12 | - 40 | - 22 | - 29 | - 21 | - 51 | |
| | | | | | | | | |

| 1 | 2 | 3 | 4 | 5 | 6 | 7 | 8 | |
|---|---|---|---|---|---|---|---|---|
| 05 | 30 | 75 | 30 | 63 | 50 | 45 | 34 | 12:2 |
| - 02 | 10 | - 12 | 20 | - 20 | 05 | 50 | 21 | |
| 60 | 05 | - 11 | 05 | 02 | - 12 | - 22 | 14 | |
| - 20 | - 21 | - 12 | - 22 | - 31 | 25 | - 62 | - 21 | |
| | | | | | | | | |

| 1 | 2 | 3 | 4 | 5 | 6 | 7 | 8 | |
|---|---|---|---|---|---|---|---|---|
| 30 | 05 | 15 | 55 | 50 | 65 | 56 | 46 | 12:3 |
| 15 | 60 | 50 | - 20 | 06 | - 21 | - 22 | 10 | |
| - 22 | - 22 | - 02 | 14 | - 22 | - 44 | 50 | - 12 | |
| - 03 | 15 | 20 | - 48 | 15 | 78 | - 73 | - 32 | |
| | | | | | | | | |

| 1 | 2 | 3 | 4 | 5 | 6 | 7 | 8 | |
|---|---|---|---|---|---|---|---|---|
| 50 | 55 | 35 | 55 | 64 | 46 | 55 | 49 | 12:4 |
| 16 | - 21 | 20 | - 12 | 01 | - 20 | - 11 | 20 | |
| - 25 | 52 | - 11 | 50 | - 22 | 21 | 12 | - 14 | |
| 17 | - 66 | 52 | 05 | - 33 | - 05 | - 25 | - 22 | |
| | | | | | | | | |

## DAY 2 – TUESDAY

TIME: _____min _____sec   Accuracy _____/16 ☆

| 1 | 2 | 3 | 4 | 5 | 6 | 7 | 8 | |
|---|---|---|---|---|---|---|---|---|
| 57<br>- 21<br>- 11<br>- 25 | 24<br>21<br>21<br>- 22 | 65<br>- 22<br>05<br>- 36 | 55<br>33<br>- 22<br>10 | 45<br>20<br>- 22<br>- 42 | 35<br>21<br>- 22<br>- 03 | 15<br>60<br>- 22<br>- 12 | 14<br>51<br>- 22<br>- 20 | 12:5 |
| | | | | | | | | |

| 1 | 2 | 3 | 4 | 5 | 6 | 7 | 8 | |
|---|---|---|---|---|---|---|---|---|
| 73<br>21<br>- 33<br>- 21 | 15<br>30<br>- 02<br>11 | 96<br>- 30<br>- 21<br>- 41 | 88<br>- 11<br>- 11<br>- 22 | 34<br>12<br>12<br>- 26 | 76<br>- 21<br>- 12<br>- 33 | 52<br>15<br>- 26<br>17 | 24<br>22<br>13<br>- 27 | 12:6 |
| | | | | | | | | |

## DAY 3 – WEDNESDAY

TIME: _____min _____sec   Accuracy _____/16 ☆

| 1 | 2 | 3 | 4 | 5 | 6 | 7 | 8 | |
|---|---|---|---|---|---|---|---|---|
| 75<br>- 21<br>- 24<br>68 | 56<br>- 22<br>51<br>- 72 | 06<br>50<br>- 22<br>- 13 | 50<br>16<br>- 22<br>15 | 65<br>- 22<br>- 31<br>- 11 | 47<br>12<br>- 24<br>- 32 | 85<br>- 12<br>25<br>- 68 | 44<br>50<br>- 22<br>- 51 | 12:7 |
| | | | | | | | | |

| 1 | 2 | 3 | 4 | 5 | 6 | 7 | 8 | |
|---|---|---|---|---|---|---|---|---|
| 67<br>- 21<br>- 15<br>27 | 66<br>- 22<br>55<br>- 77 | 55<br>- 11<br>21<br>- 11 | 34<br>21<br>31<br>13 | 44<br>21<br>01<br>- 25 | 24<br>21<br>- 30<br>- 12 | 43<br>25<br>- 10<br>- 25 | 96<br>- 42<br>- 12<br>27 | 12:8 |
| | | | | | | | | |

## DAY 4 – THURSDAY

TIME: _____min _____sec   Accuracy _____/16 ☆

| 1 | 2 | 3 | 4 | 5 | 6 | 7 | 8 |
|---|---|---|---|---|---|---|---|
| 15 | 75 | 43 | 44 | 64 | 54 | 65 | 43 |
| - 11 | - 21 | 12 | 12 | - 21 | 11 | - 21 | 12 |
| 20 | 01 | - 21 | - 21 | 22 | - 20 | - 01 | - 22 |
| 51 | 04 | 50 | 03 | - 10 | - 01 | 52 | 50 |
|  |  |  |  |  |  |  |  |

12:9

| 1 | 2 | 3 | 4 | 5 | 6 | 7 | 8 |
|---|---|---|---|---|---|---|---|
| 34 | 50 | 45 | 74 | 11 | 35 | 34 | 36 |
| 01 | 05 | 11 | 02 | 23 | - 21 | 22 | 13 |
| 02 | - 50 | 13 | - 66 | 22 | 71 | - 21 | - 44 |
| 22 | 44 | - 27 | 18 | - 10 | 04 | - 31 | - 02 |
|  |  |  |  |  |  |  |  |

12:10

## DAY 5 – FRIDAY

TIME: _____min _____sec   Accuracy _____/16 ☆

| 1 | 2 | 3 | 4 | 5 | 6 | 7 | 8 |
|---|---|---|---|---|---|---|---|
| 34 | 45 | 44 | 36 | 31 | 53 | 54 | 45 |
| - 22 | - 12 | 21 | - 02 | 25 | - 21 | 02 | 24 |
| 52 | 21 | - 22 | 21 | - 12 | 02 | - 22 | - 21 |
| 12 | - 20 | 12 | 21 | 21 | 20 | 10 | - 11 |
|  |  |  |  |  |  |  |  |

12:11

| 1 | 2 | 3 | 4 | 5 | 6 | 7 | 8 |
|---|---|---|---|---|---|---|---|
| 94 | 33 | 87 | 54 | 25 | 52 | 61 | 23 |
| - 42 | 52 | - 15 | 15 | 71 | 11 | 05 | 21 |
| - 20 | - 31 | 02 | - 26 | - 42 | - 20 | - 21 | 22 |
| 56 | - 24 | 01 | 15 | - 10 | 26 | - 32 | - 55 |
|  |  |  |  |  |  |  |  |

12:12

www.abacus-math.com

## LESSON 12 – MIND MATH PRACTICE

### DAY 1 – MONDAY          Accuracy _____/10 ☆

| 1 | 2 | 3 | 4 | 5 | 6 | 7 | 8 | 9 | 10 |
|---|---|---|---|---|---|---|---|---|---|
|    |    |    |    |    | 39 | 50 | 04 | 50 | 45 |
| 04 | 40 | 05 | 45 | 07 | - 05 | - 20 | 01 | 10 | - 02 |
| 01 | 10 | - 02 | - 02 | - 02 | 10 | 10 | - 02 | - 20 | 10 |
| - 02 | 07 | - 02 | - 01 | - 02 | - 20 | 10 | - 02 | - 20 | - 20 |
|    |    |    |    |    |    |    |    |    |    |

12:13

### DAY 2 – TUESDAY          Accuracy _____/10 ☆

| 1 | 2 | 3 | 4 | 5 | 6 | 7 | 8 | 9 | 10 |
|---|---|---|---|---|---|---|---|---|---|
|    |    |    |    |    | 25 | 33 | 34 | 45 | 46 |
| 35 | 24 | 25 | 44 | 14 | - 11 | 02 | 10 | 12 | 10 |
| 02 | 21 | - 02 | 11 | 11 | 01 | - 01 | 10 | - 21 | - 20 |
| - 22 | 10 | - 20 | - 02 | - 02 | - 02 | - 01 | - 20 | - 11 | - 02 |
|    |    |    |    |    |    |    |    |    |    |

12:14

### DAY 3 – WEDNESDAY          Accuracy _____/10 ☆

| 1 | 2 | 3 | 4 | 5 | 6 | 7 | 8 | 9 | 10 |
|---|---|---|---|---|---|---|---|---|---|
| 30 | 54 | 54 | 60 | 35 | 43 | 64 | 53 | 64 | 53 |
| 20 | - 20 | 02 | - 20 | 20 | 11 | - 23 | - 21 | - 22 | - 22 |
| - 10 | - 10 | - 20 | 10 | - 01 | - 13 | - 11 | - 22 | 16 | - 11 |
| 20 | 01 | - 10 | - 20 | - 20 | 10 | 26 | - 10 | - 18 | 69 |
|    |    |    |    |    |    |    |    |    |    |

12:15

### DAY 4 – THURSDAY          Accuracy _____/10 ☆

| 1 | 2 | 3 | 4 | 5 | 6 | 7 | 8 | 9 | 10 |
|---|---|---|---|---|---|---|---|---|---|
| 60 | 50 | 55 | 55 | 34 | 15 | 54 | 34 | 43 | 74 |
| 05 | 04 | - 11 | - 02 | 20 | - 11 | 12 | 21 | 22 | 01 |
| 04 | 02 | 55 | 01 | 01 | 01 | - 11 | 33 | - 01 | - 20 |
| - 29 | - 20 | - 71 | 40 | - 12 | - 02 | - 25 | - 57 | - 24 | - 12 |
|    |    |    |    |    |    |    |    |    |    |

12:16

| 1 | 2 | 3 | 4 | 5 | 6 | 7 | 8 | 9 | 10 |
|---|---|---|---|---|---|---|---|---|----|
| 21 | 85 | 33 | 24 | 54 | 36 | 55 | 54 | 85 | 65 |
| 22 | - 32 | 11 | 52 | 42 | - 22 | - 10 | 01 | - 71 | 10 |
| 10 | - 20 | 11 | - 20 | 01 | 51 | 50 | - 20 | 62 | - 51 |
| - 20 | - 11 | - 22 | - 02 | - 56 | - 15 | - 02 | - 02 | - 20 | - 02 |
|  |  |  |  |  |  |  |  |  |  |

12:17  © SAI Speed Math Academy, USA

| 1 | 2 | 3 | 4 | 5 | 6 | 7 | 8 | 9 | 10 |
|---|---|---|---|---|---|---|---|---|----|
| 64 | 55 | 57 | 55 | 40 | 36 | 89 | 51 | 96 | 64 |
| - 21 | - 12 | - 22 | - 21 | 10 | - 01 | - 56 | 03 | - 72 | 01 |
| - 42 | - 22 | - 02 | 11 | 05 | 60 | 25 | 41 | 61 | - 21 |
| - 01 | - 21 | 56 | - 25 | - 22 | - 52 | - 18 | 04 | - 82 | - 44 |
|  |  |  |  |  |  |  |  |  |  |

12:18  © SAI Speed Math Academy, USA

## SHAPE PUZZLE

Find out what each shape is worth.

Suggestion: Take as many numbers of beans, blocks or coins as the right side of the equation shows and share them on the shapes to the left side of the equation.

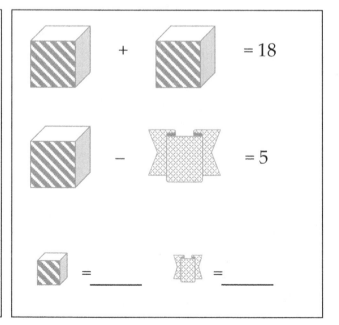

## LESSON 12 – DICTATION

**DO 6 PROBLEMS A DAY** and write answers below.

Students: Use abacus to calculate dictation problems this week.

Teachers: Dictate problems from mind math part of this week's homework.

| 1 | 2 | 3 | 4 | 5 | 6 | 7 | 8 | 9 | 10 |
|---|---|---|---|---|---|---|---|---|----|
|   |   |   |   |   |   |   |   |   |    |

| 11 | 12 | 13 | 14 | 15 | 16 | 17 | 18 | 19 | 20 |
|----|----|----|----|----|----|----|----|----|----|
|    |    |    |    |    |    |    |    |    |    |

| 21 | 22 | 23 | 24 | 25 | 26 | 27 | 28 | 29 | 30 |
|----|----|----|----|----|----|----|----|----|----|
|    |    |    |    |    |    |    |    |    |    |

## LESSON 12 – SKILL BUILDING

Visualize the numbers on the beam in your mind and draw it to represent the numbers given.

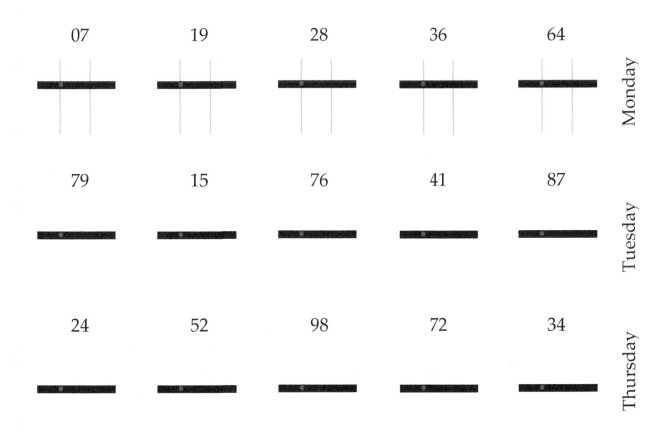

| 07 | 19 | 28 | 36 | 64 | Monday |

| 79 | 15 | 76 | 41 | 87 | Tuesday |

| 24 | 52 | 98 | 72 | 34 | Thursday |

# WEEK 13 – LESSON 13 – INTRODUCING +3 CONCEPT

## LESSON 13 – PRACTICE WORK

+3 = + 5 – 2

Accuracy _____ /32  ☆

Use Abacus

### DAY 1 – MONDAY

| 1 | 2 | 3 | 4 | 5 | 6 | 7 | 8 | |
|---|---|---|---|---|---|---|---|---|
| 21 | 44 | 14 | 43 | 44 | 24 | 32 | 42 | 13:1 |
| 03 | 30 | 23 | - 03 | 23 | 22 | 33 | 03 | |
| 23 | 03 | 10 | 39 | 32 | 33 | - 05 | 20 | |
| 12 | - 05 | 30 | - 16 | - 11 | - 20 | 01 | 20 | |
| | | | | | | | | |

| 1 | 2 | 3 | 4 | 5 | 6 | 7 | 8 | |
|---|---|---|---|---|---|---|---|---|
| 14 | 45 | 49 | 55 | 54 | 34 | 35 | 46 | 13:2 |
| 32 | 34 | - 15 | - 21 | 21 | 33 | - 21 | 32 | |
| 30 | - 57 | - 32 | 32 | - 12 | - 51 | 63 | - 15 | |
| 02 | 33 | 13 | - 11 | 13 | - 02 | - 15 | 33 | |
| | | | | | | | | |

| 1 | 2 | 3 | 4 | 5 | 6 | 7 | 8 | |
|---|---|---|---|---|---|---|---|---|
| 33 | 31 | 24 | 22 | 33 | 13 | 22 | 21 | 13:3 |
| 23 | 11 | 23 | 10 | 33 | 12 | 23 | 32 | |
| - 50 | 33 | 30 | 13 | - 55 | 33 | 32 | 13 | |
| 43 | 02 | - 15 | 30 | 86 | 20 | - 50 | - 22 | |
| | | | | | | | | |

| 1 | 2 | 3 | 4 | 5 | 6 | 7 | 8 | |
|---|---|---|---|---|---|---|---|---|
| 23 | 65 | 43 | 75 | 45 | 54 | 45 | 13 | 13:4 |
| 13 | - 21 | 12 | - 22 | 22 | 32 | 24 | 23 | |
| 32 | 33 | - 20 | - 23 | - 11 | - 62 | - 55 | 30 | |
| - 26 | - 66 | 34 | 34 | - 52 | 30 | 83 | - 25 | |
| | | | | | | | | |

## DAY 2 – TUESDAY

TIME: _____ min _____ sec   Accuracy _____ /16

| 1 | 2 | 3 | 4 | 5 | 6 | 7 | 8 |
|---|---|---|---|---|---|---|---|
| 41 | 44 | 84 | 24 | 12 | 22 | 43 | 28 |
| 13 | 32 | - 62 | 33 | 33 | 35 | 23 | 31 |
| 23 | 20 | 32 | - 55 | 21 | - 12 | 11 | - 54 |
| - 11 | 03 | - 11 | 13 | - 60 | 33 | 20 | 13 |
|  |  |  |  |  |  |  |  |

13:5

| 1 | 2 | 3 | 4 | 5 | 6 | 7 | 8 |
|---|---|---|---|---|---|---|---|
| 32 | 42 | 34 | 25 | 62 | 87 | 99 | 68 |
| 23 | 33 | 31 | - 21 | 23 | - 72 | - 21 | - 22 |
| - 12 | - 55 | 20 | 22 | - 31 | - 11 | - 22 | 30 |
| 03 | 34 | 14 | 33 | - 23 | 23 | - 11 | - 72 |
|  |  |  |  |  |  |  |  |

13:6

## DAY 3 – WEDNESDAY

TIME: _____ min _____ sec   Accuracy _____ /16

| 1 | 2 | 3 | 4 | 5 | 6 | 7 | 8 |
|---|---|---|---|---|---|---|---|
| 25 | 54 | 34 | 43 | 46 | 55 | 55 | 52 |
| - 12 | 12 | 31 | 23 | 31 | - 21 | - 11 | 13 |
| 31 | - 11 | - 11 | - 21 | - 22 | 23 | 50 | - 20 |
| - 12 | - 21 | - 03 | - 31 | - 12 | - 12 | 03 | - 11 |
|  |  |  |  |  |  |  |  |

13:7

| 1 | 2 | 3 | 4 | 5 | 6 | 7 | 8 |
|---|---|---|---|---|---|---|---|
| 85 | 64 | 23 | 65 | 43 | 23 | 43 | 85 |
| - 72 | 13 | 32 | - 12 | 32 | 33 | 13 | - 51 |
| 13 | - 25 | - 22 | - 13 | - 22 | 31 | - 12 | - 12 |
| 13 | - 51 | 33 | 39 | 31 | - 66 | 23 | 36 |
|  |  |  |  |  |  |  |  |

13:8

## DAY 4 – THURSDAY

TIME: _____min _____sec   Accuracy _____/16   ☆

| 1 | 2 | 3 | 4 | 5 | 6 | 7 | 8 |
|---|---|---|---|---|---|---|---|
| 45 | 65 | 24 | 44 | 66 | 76 | 85 | 43 |
| - 10 | - 21 | 21 | 31 | - 21 | - 21 | - 32 | 13 |
| 31 | 13 | - 32 | - 21 | 32 | - 11 | - 11 | - 22 |
| - 51 | - 15 | 15 | 33 | - 11 | 31 | 03 | - 24 |
| | | | | | | | |

| 1 | 2 | 3 | 4 | 5 | 6 | 7 | 8 |
|---|---|---|---|---|---|---|---|
| 34 | 51 | 53 | 26 | 25 | 99 | 20 | 29 |
| 21 | 25 | 12 | 23 | 32 | - 13 | 34 | 30 |
| - 02 | - 22 | - 25 | 30 | - 16 | - 32 | 23 | - 14 |
| 23 | - 14 | 39 | - 25 | 35 | - 14 | - 77 | - 42 |
| | | | | | | | |

## DAY 5 – FRIDAY

TIME: _____min _____sec   Accuracy _____/16   ☆

| 1 | 2 | 3 | 4 | 5 | 6 | 7 | 8 |
|---|---|---|---|---|---|---|---|
| 14 | 23 | 32 | 21 | 15 | 55 | 75 | 53 |
| 13 | - 11 | - 21 | 35 | - 11 | - 11 | - 21 | 12 |
| - 02 | 23 | 51 | 10 | 43 | 33 | 13 | - 22 |
| - 01 | - 31 | - 22 | - 52 | - 32 | - 21 | - 12 | - 32 |
| | | | | | | | |

| 1 | 2 | 3 | 4 | 5 | 6 | 7 | 8 |
|---|---|---|---|---|---|---|---|
| 35 | 46 | 43 | 42 | 75 | 66 | 67 | 11 |
| 31 | 32 | 13 | 23 | - 52 | 22 | - 21 | 32 |
| - 21 | - 56 | - 25 | - 11 | 35 | - 53 | 33 | 23 |
| - 31 | 32 | - 31 | - 21 | - 17 | 34 | - 09 | - 65 |
| | | | | | | | |

www.abacus-math.com

# LESSON 13 – MIND MATH PRACTICE

 Visualize

## DAY 1 – MONDAY          Accuracy _____/10 ☆

| 1 | 2 | 3 | 4 | 5 | 6 | 7 | 8 | 9 | 10 |
|---|---|---|---|---|---|---|---|---|---|
|  |  |  |  |  | 19 | 55 | 14 | 24 | 35 |
| 14 | 03 | 13 | 22 | 25 | - 05 | - 11 | 11 | 23 | - 22 |
| 03 | 03 | 13 | 30 | 30 | 03 | 02 | 30 | 30 | 11 |
| - 06 | - 05 | - 20 | - 10 | - 55 | 30 | 30 | - 02 | - 50 | 30 |
|  |  |  |  |  |  |  |  |  |  |

13:13

## DAY 2 – TUESDAY          Accuracy _____/10 ☆

| 1 | 2 | 3 | 4 | 5 | 6 | 7 | 8 | 9 | 10 |
|---|---|---|---|---|---|---|---|---|---|
|  |  |  |  |  | 25 | 33 | 24 | 34 | 46 |
| 62 | 03 | 35 | 41 | 25 | - 11 | 13 | 32 | 30 | 10 |
| 03 | 03 | - 01 | 30 | 30 | 03 | - 01 | 20 | - 10 | - 20 |
| - 01 | 20 | 03 | - 21 | 10 | 30 | - 01 | - 55 | - 20 | - 12 |
|  |  |  |  |  |  |  |  |  |  |

13:14

## DAY 3 – WEDNESDAY          Accuracy _____/10 ☆

| 1 | 2 | 3 | 4 | 5 | 6 | 7 | 8 | 9 | 10 |
|---|---|---|---|---|---|---|---|---|---|
| 04 | 34 | 34 | 74 | 34 | 23 | 13 | 20 | 31 | 44 |
| 03 | 03 | 13 | - 23 | 30 | 13 | 31 | 31 | 33 | - 03 |
| - 01 | - 20 | 30 | - 10 | - 50 | - 20 | 55 | 31 | - 02 | - 21 |
| 03 | - 12 | - 22 | 30 | 03 | - 11 | - 83 | 03 | 03 | 35 |
|  |  |  |  |  |  |  |  |  |  |

13:15

## DAY 4 – THURSDAY          Accuracy _____/10 ☆

| 1 | 2 | 3 | 4 | 5 | 6 | 7 | 8 | 9 | 10 |
|---|---|---|---|---|---|---|---|---|---|
| 79 | 55 | 35 | 43 | 44 | 55 | 32 | 23 | 43 | 74 |
| - 23 | - 12 | - 20 | 33 | 11 | - 20 | 35 | 31 | 13 | - 21 |
| - 01 | 30 | 30 | 20 | - 22 | - 01 | - 56 | - 20 | - 02 | 33 |
| - 02 | - 53 | 30 | - 60 | 33 | 03 | 88 | 30 | - 02 | - 21 |
|  |  |  |  |  |  |  |  |  |  |

13:16

| 1 | 2 | 3 | 4 | 5 | 6 | 7 | 8 | 9 | 10 | |
|---|---|---|---|---|---|---|---|---|---|---|
| 43 | 85 | 33 | 04 | 11 | 43 | 42 | 44 | 24 | 64 | |
| 03 | - 32 | 03 | 53 | 13 | 13 | 31 | 30 | 23 | - 20 | 13:17 |
| - 11 | - 11 | - 11 | - 10 | 13 | - 20 | - 50 | - 51 | - 11 | 30 | |
| 30 | 03 | 30 | - 12 | - 22 | 30 | 31 | 30 | 03 | 03 | |
|  |  |  |  |  |  |  |  |  |  | |

| 1 | 2 | 3 | 4 | 5 | 6 | 7 | 8 | 9 | 10 | |
|---|---|---|---|---|---|---|---|---|---|---|
| 36 | 54 | 51 | 75 | 54 | 34 | 41 | 40 | 13 | 45 | |
| - 22 | - 01 | 03 | - 61 | 03 | 33 | - 30 | 33 | 63 | 32 | 13:18 |
| 03 | 03 | - 02 | 33 | - 52 | - 20 | 23 | - 11 | - 25 | 10 | |
| - 11 | - 02 | 03 | - 02 | - 01 | 30 | 03 | 03 | - 11 | - 55 | |
|  |  |  |  |  |  |  |  |  |  | |

## SHAPE PUZZLE

Find out what each shape is worth.   Color the shapes.

Suggestion:  Take as many numbers of beans, blocks or coins as the right side of the equation shows and share them on the shapes to the left side of the equation.

## LESSON 13 – DICTATION

**DO 6 PROBLEMS A DAY** and write answers below.

Students: Every day use abacus to calculate three dictation problems and try mind math for three dictation problems.

Teachers: Dictate problems from mind math part of this week's homework.

| 1 | 2 | 3 | 4 | 5 | 6 | 7 | 8 | 9 | 10 |
|---|---|---|---|---|---|---|---|---|---|
|   |   |   |   |   |   |   |   |   |   |

| 11 | 12 | 13 | 14 | 15 | 16 | 17 | 18 | 19 | 20 |
|---|---|---|---|---|---|---|---|---|---|
|   |   |   |   |   |   |   |   |   |   |

| 21 | 22 | 23 | 24 | 25 | 26 | 27 | 28 | 29 | 30 |
|---|---|---|---|---|---|---|---|---|---|
|   |   |   |   |   |   |   |   |   |   |

## LESSON 13 – SKILL BUILDING

Visualize the numbers on the beam in your mind and draw it to represent the numbers given.

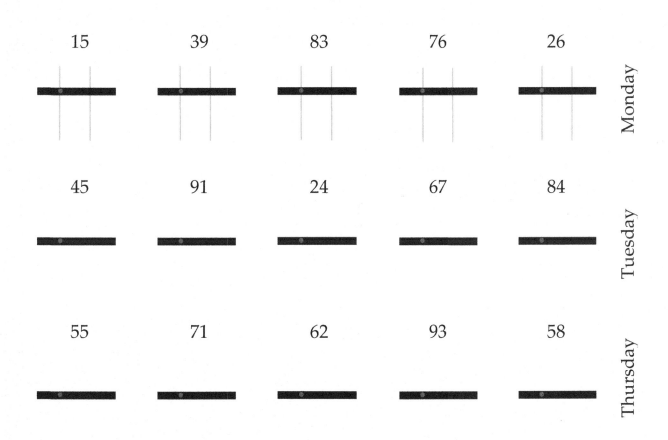

| 15 | 39 | 83 | 76 | 26 | Monday |
| 45 | 91 | 24 | 67 | 84 | Tuesday |
| 55 | 71 | 62 | 93 | 58 | Thursday |

# WEEK 14 – LESSON 14 – INTRODUCING –3 CONCEPT

## LESSON 14 – PRACTICE WORK

$$-3 = -5 + 2$$

**DAY 1 – MONDAY**　　　Accuracy _____/32　☆　　| Use Abacus ▨ |

| 1 | 2 | 3 | 4 | 5 | 6 | 7 | 8 | |
|---|---|---|---|---|---|---|---|---|
| 25 | 54 | 55 | 50 | 35 | 31 | 20 | 55 | 14:1 |
| 10 | - 33 | 13 | - 30 | 22 | 15 | 23 | - 33 | |
| - 03 | 11 | - 30 | 10 | - 31 | 10 | 01 | 66 | |
| 20 | 21 | - 10 | - 30 | - 21 | - 33 | - 33 | - 22 | |
| | | | | | | | | |

| 1 | 2 | 3 | 4 | 5 | 6 | 7 | 8 | |
|---|---|---|---|---|---|---|---|---|
| 57 | 32 | 45 | 65 | 24 | 44 | 66 | 46 | 14:2 |
| - 32 | 33 | - 13 | - 31 | 21 | 31 | - 23 | 12 | |
| 21 | 12 | 31 | 11 | - 32 | - 21 | 32 | - 33 | |
| - 32 | - 23 | - 51 | 14 | 15 | 33 | - 13 | - 23 | |
| | | | | | | | | |

| 1 | 2 | 3 | 4 | 5 | 6 | 7 | 8 | |
|---|---|---|---|---|---|---|---|---|
| 75 | 30 | 60 | 33 | 42 | 31 | 15 | 16 | 14:3 |
| - 33 | 24 | - 30 | 22 | 12 | 25 | 60 | 53 | |
| - 11 | 22 | 25 | - 31 | - 33 | - 33 | - 33 | - 14 | |
| 25 | - 30 | - 33 | 55 | - 11 | - 03 | 21 | - 31 | |
| | | | | | | | | |

| 1 | 2 | 3 | 4 | 5 | 6 | 7 | 8 | |
|---|---|---|---|---|---|---|---|---|
| 35 | 23 | 54 | 85 | 43 | 34 | 51 | 37 | 14:4 |
| - 21 | 52 | 21 | - 32 | 13 | 21 | 25 | 22 | |
| 12 | - 31 | - 32 | - 31 | - 22 | - 33 | - 32 | - 33 | |
| - 03 | 15 | - 32 | 53 | 60 | 52 | - 44 | - 03 | |
| | | | | | | | | |

## DAY 2 – TUESDAY

TIME: _____ min _____ sec   Accuracy _____ /16 ☆

| 1 | 2 | 3 | 4 | 5 | 6 | 7 | 8 | |
|---|---|---|---|---|---|---|---|---|
| 49 | 55 | 54 | 34 | 35 | 43 | 75 | 45 | |
| - 15 | - 21 | 21 | 21 | - 23 | 12 | - 32 | 22 | |
| - 32 | 12 | - 30 | - 53 | 61 | - 23 | - 23 | - 31 | 14:5 |
| 13 | - 43 | - 13 | 06 | - 32 | 32 | 34 | 22 | |
| | | | | | | | | |

| 1 | 2 | 3 | 4 | 5 | 6 | 7 | 8 | |
|---|---|---|---|---|---|---|---|---|
| 54 | 45 | 65 | 56 | 16 | 87 | 55 | 68 | |
| 32 | 24 | - 31 | - 32 | 51 | - 13 | - 03 | - 23 | |
| - 21 | - 38 | - 34 | 51 | - 23 | - 33 | 11 | - 31 | 14:6 |
| - 33 | 22 | 87 | - 73 | - 32 | 15 | - 23 | 75 | |
| | | | | | | | | |

## DAY 3 – WEDNESDAY

TIME: _____ min _____ sec   Accuracy _____ /16 ☆

| 1 | 2 | 3 | 4 | 5 | 6 | 7 | 8 | |
|---|---|---|---|---|---|---|---|---|
| 96 | 54 | 68 | 43 | 46 | 76 | 55 | 52 | |
| - 33 | 12 | 21 | 23 | 11 | - 12 | - 11 | 13 | |
| - 12 | - 11 | - 55 | - 21 | - 32 | - 31 | 50 | 21 | 14:7 |
| - 30 | - 31 | 11 | - 13 | - 21 | - 13 | 01 | - 53 | |
| | | | | | | | | |

| 1 | 2 | 3 | 4 | 5 | 6 | 7 | 8 | |
|---|---|---|---|---|---|---|---|---|
| 85 | 64 | 34 | 68 | 82 | 46 | 95 | 96 | |
| - 73 | 13 | 22 | - 31 | 15 | - 30 | - 31 | - 53 | |
| 61 | - 35 | - 21 | 22 | - 22 | 21 | - 30 | 12 | 14:8 |
| 13 | - 41 | - 13 | - 35 | - 33 | - 03 | - 34 | - 35 | |
| | | | | | | | | |

## DAY 4 – THURSDAY

TIME: _____min _____sec    Accuracy _____/16  ☆

| 1 | 2 | 3 | 4 | 5 | 6 | 7 | 8 |
|---|---|---|---|---|---|---|---|
| 11 | 52 | 44 | 57 | 58 | 44 | 58 | 67 |
| 23 | - 31 | 23 | - 13 | - 23 | 31 | - 22 | - 23 |
| 13 | 23 | - 31 | 51 | - 23 | - 23 | - 13 | 11 |
| - 36 | 23 | - 15 | - 22 | - 11 | - 32 | 35 | - 23 |
|  |  |  |  |  |  |  |  |

14:9

| 1 | 2 | 3 | 4 | 5 | 6 | 7 | 8 |
|---|---|---|---|---|---|---|---|
| 34 | 43 | 51 | 74 | 15 | 62 | 28 | 87 |
| 33 | 03 | 12 | 15 | 61 | - 21 | 20 | 11 |
| - 22 | - 31 | 12 | - 33 | - 53 | 55 | - 42 | - 33 |
| - 43 | 54 | - 73 | - 33 | - 10 | - 43 | - 03 | - 31 |
|  |  |  |  |  |  |  |  |

14:10

## DAY 5 – FRIDAY

TIME: _____min _____sec    Accuracy _____/16  ☆

| 1 | 2 | 3 | 4 | 5 | 6 | 7 | 8 |
|---|---|---|---|---|---|---|---|
| 24 | 85 | 76 | 52 | 63 | 25 | 99 | 36 |
| 31 | - 62 | - 23 | 13 | 24 | 31 | - 21 | 31 |
| - 13 | 35 | 32 | - 30 | - 33 | 11 | - 32 | - 27 |
| 15 | - 53 | - 83 | - 21 | - 54 | - 23 | - 32 | 55 |
|  |  |  |  |  |  |  |  |

14:11

| 1 | 2 | 3 | 4 | 5 | 6 | 7 | 8 |
|---|---|---|---|---|---|---|---|
| 58 | 25 | 54 | 58 | 48 | 34 | 75 | 47 |
| 31 | 32 | 12 | 20 | 51 | 22 | - 21 | 32 |
| - 26 | - 13 | - 35 | - 33 | - 35 | - 13 | - 34 | - 67 |
| 15 | 50 | 61 | - 41 | 11 | - 30 | - 20 | 13 |
|  |  |  |  |  |  |  |  |

14:12

# LESSON 14 – MIND MATH PRACTICE

## DAY 1 – MONDAY

Accuracy _____/10 ☆

| 1 | 2 | 3 | 4 | 5 | 6 | 7 | 8 | 9 | 10 |
|---|---|---|---|---|---|---|---|---|---|
| | | | | | 19 | 55 | 14 | 24 | 35 |
| 35 | 16 | 17 | 50 | 68 | - 04 | - 11 | 11 | 22 | - 03 |
| - 03 | - 03 | - 03 | - 30 | - 33 | 22 | 20 | - 13 | - 03 | 11 |
| - 02 | 06 | - 11 | 70 | - 05 | - 30 | - 34 | 22 | - 03 | - 33 |
| | | | | | | | | | |

14:13

© SAI Speed Math Academy, USA

## DAY 2 – TUESDAY

Accuracy _____/10 ☆

| 1 | 2 | 3 | 4 | 5 | 6 | 7 | 8 | 9 | 10 |
|---|---|---|---|---|---|---|---|---|---|
| | | | | | 25 | 43 | 24 | 34 | 46 |
| 62 | 45 | 80 | 27 | 51 | - 03 | 10 | 10 | 20 | 10 |
| 02 | 21 | - 30 | 10 | - 30 | 02 | - 30 | 20 | - 32 | - 30 |
| - 33 | - 03 | - 30 | - 03 | 08 | 02 | - 20 | - 30 | - 21 | - 13 |
| | | | | | | | | | |

14:14

© SAI Speed Math Academy, USA

## DAY 3 – WEDNESDAY

Accuracy _____/10 ☆

| 1 | 2 | 3 | 4 | 5 | 6 | 7 | 8 | 9 | 10 |
|---|---|---|---|---|---|---|---|---|---|
| 04 | 34 | 44 | 74 | 45 | 23 | 13 | 55 | 31 | 66 |
| 02 | 11 | 12 | - 30 | - 03 | 12 | 31 | - 33 | 23 | - 03 |
| - 03 | - 30 | - 03 | - 31 | 10 | - 03 | 55 | 05 | - 32 | - 30 |
| 02 | - 13 | - 20 | 70 | - 30 | - 11 | - 83 | - 03 | 12 | - 33 |
| | | | | | | | | | |

14:15

© SAI Speed Math Academy, USA

## DAY 4 – THURSDAY

Accuracy _____/10 ☆

| 1 | 2 | 3 | 4 | 5 | 6 | 7 | 8 | 9 | 10 |
|---|---|---|---|---|---|---|---|---|---|
| 94 | 55 | 35 | 63 | 44 | 55 | 32 | 23 | 66 | 77 |
| - 03 | - 13 | - 23 | - 33 | 22 | - 30 | 25 | 31 | 12 | - 23 |
| - 41 | 20 | 11 | 25 | - 33 | - 03 | - 36 | - 20 | - 30 | - 30 |
| - 30 | - 30 | 01 | - 03 | 11 | 21 | 60 | - 34 | - 40 | 01 |
| | | | | | | | | | |

14:16

© SAI Speed Math Academy, USA

Accuracy _____/20 ☆

| 1 | 2 | 3 | 4 | 5 | 6 | 7 | 8 | 9 | 10 | |
|---|---|---|---|---|---|---|---|---|---|---|
| 43 | 65 | 33 | 14 | 27 | 36 | 54 | 51 | 75 | 64 | |
| 20 | - 30 | 02 | 52 | - 12 | - 03 | - 31 | 03 | - 61 | 11 | |
| - 30 | - 20 | - 13 | - 30 | - 13 | 20 | 10 | - 31 | 11 | - 53 | 14:17 |
| - 11 | - 13 | 22 | - 03 | - 02 | - 31 | - 03 | 02 | - 03 | - 01 | |
|  |  |  |  |  |  |  |  |  |  | |

| 1 | 2 | 3 | 4 | 5 | 6 | 7 | 8 | 9 | 10 | |
|---|---|---|---|---|---|---|---|---|---|---|
| 56 | 42 | 44 | 54 | 86 | 34 | 41 | 40 | 13 | 45 | |
| 01 | 21 | 20 | 21 | - 03 | 12 | - 30 | 22 | 62 | 11 | |
| - 30 | - 30 | - 51 | - 03 | 11 | - 03 | 23 | - 31 | - 05 | 10 | 14:18 |
| - 23 | 55 | 33 | 05 | - 44 | 55 | 20 | 20 | - 30 | - 35 | |
|  |  |  |  |  |  |  |  |  |  | |

## SHAPE PUZZLE

Find out what each shape is worth.   Color the shapes.

Suggestion: Take as many numbers of beans, blocks or coins as the right side of the equation shows and share them on the shapes to the left side of the equation.

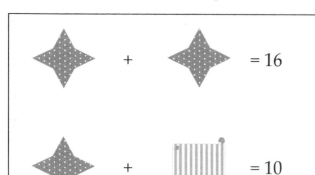

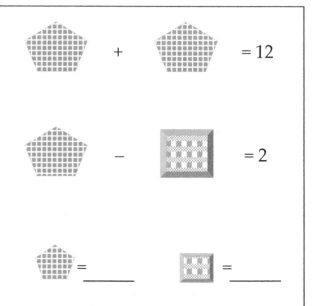

## LESSON 14 – DICTATION

**DO 6 PROBLEMS A DAY** and write answers below.

Students: Every day use abacus to calculate three dictation problems and try mind math for three dictation problems.

Teachers: Dictate any problem from this week's homework.

| 1 | 2 | 3 | 4 | 5 | 6 | 7 | 8 | 9 | 10 |
|---|---|---|---|---|---|---|---|---|---|
| | | | | | | | | | |

| 11 | 12 | 13 | 14 | 15 | 16 | 17 | 18 | 19 | 20 |
|---|---|---|---|---|---|---|---|---|---|
| | | | | | | | | | |

| 21 | 22 | 23 | 24 | 25 | 26 | 27 | 28 | 29 | 30 |
|---|---|---|---|---|---|---|---|---|---|
| | | | | | | | | | |

## LESSON 14 – SKILL BUILDING

Visualize the numbers on the beam in your mind and draw it to represent the numbers given.

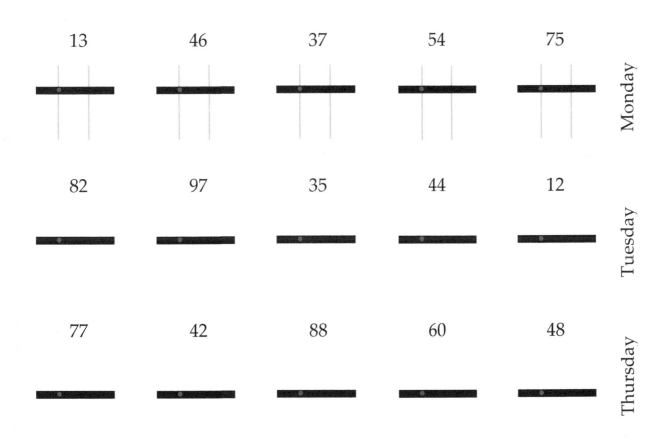

| 13 | 46 | 37 | 54 | 75 | Monday |

| 82 | 97 | 35 | 44 | 12 | Tuesday |

| 77 | 42 | 88 | 60 | 48 | Thursday |

# WEEK 15 – SKILL BUILDING

## WEEK 15 – PRACTICE WORK

Use Abacus

### DAY 1 – MONDAY

TIME: _____min _____sec   Accuracy _____/16 ⭐

| 1 | 2 | 3 | 4 | 5 | 6 | 7 | 8 |
|---|---|---|---|---|---|---|---|
| 53 | 15 | 44 | 11 | 34 | 34 | 51 | 55 |
| - 31 | 30 | 21 | 22 | 01 | 21 | 25 | - 03 |
| 55 | 54 | - 31 | 22 | - 03 | - 02 | - 32 | 31 |
| - 23 | - 89 | - 13 | - 33 | 20 | 23 | - 44 | - 12 |
|  |  |  |  |  |  |  |  |

15:1

| 1 | 2 | 3 | 4 | 5 | 6 | 7 | 8 |
|---|---|---|---|---|---|---|---|
| 44 | 45 | 53 | 55 | 62 | 45 | 52 | 39 |
| 10 | 12 | 21 | - 13 | 02 | - 13 | 15 | 30 |
| - 30 | - 13 | - 13 | 22 | - 32 | 52 | - 33 | - 15 |
| 20 | - 33 | - 31 | - 31 | - 32 | - 30 | 12 | - 30 |
|  |  |  |  |  |  |  |  |

15:2

### DAY 2 – TUESDAY

TIME: _____min _____sec   Accuracy _____/16 ⭐

| 1 | 2 | 3 | 4 | 5 | 6 | 7 | 8 |
|---|---|---|---|---|---|---|---|
| 54 | 85 | 66 | 35 | 43 | 24 | 86 | 66 |
| 21 | - 32 | - 13 | 22 | 11 | 21 | - 31 | - 23 |
| - 32 | - 30 | 22 | - 23 | - 33 | - 32 | - 31 | 32 |
| - 32 | 21 | - 31 | - 34 | 11 | 15 | 53 | - 11 |
|  |  |  |  |  |  |  |  |

15:3

| 1 | 2 | 3 | 4 | 5 | 6 | 7 | 8 |
|---|---|---|---|---|---|---|---|
| 43 | 34 | 76 | 51 | 14 | 65 | 62 | 68 |
| 13 | 51 | - 13 | 25 | 51 | - 33 | 23 | - 35 |
| - 22 | - 30 | 22 | - 30 | - 23 | 22 | - 31 | 13 |
| 60 | - 03 | - 71 | - 41 | 11 | 33 | - 34 | - 45 |
|  |  |  |  |  |  |  |  |

15:4

www.abacus-math.com

## DAY 3 – WEDNESDAY

TIME: _____ min _____ sec    Accuracy _____ /16 ☆

| 1 | 2 | 3 | 4 | 5 | 6 | 7 | 8 | |
|---|---|---|---|---|---|---|---|---|
| 89 | 99 | 24 | 54 | 45 | 65 | 56 | 29 | |
| - 23 | - 64 | 31 | 10 | - 13 | - 30 | - 13 | 30 | |
| - 13 | 23 | - 23 | - 33 | 21 | 11 | 51 | - 04 | 15:5 |
| - 23 | - 18 | 37 | 11 | - 51 | 12 | - 94 | - 33 | |
| | | | | | | | | |

| 1 | 2 | 3 | 4 | 5 | 6 | 7 | 8 | |
|---|---|---|---|---|---|---|---|---|
| 34 | 87 | 68 | 32 | 55 | 52 | 85 | 85 | |
| 11 | - 31 | - 23 | 57 | - 13 | 12 | - 73 | - 63 | |
| 32 | - 21 | 32 | - 64 | 50 | 21 | 61 | 12 | 15:6 |
| - 15 | - 11 | - 66 | 51 | 01 | 11 | 11 | 11 | |
| | | | | | | | | |

## DAY 4 – THURSDAY

TIME: _____ min _____ sec    Accuracy _____ /16 ☆

| 1 | 2 | 3 | 4 | 5 | 6 | 7 | 8 | |
|---|---|---|---|---|---|---|---|---|
| 76 | 54 | 55 | 95 | 25 | 64 | 54 | 65 | |
| - 31 | 12 | 22 | - 61 | - 13 | 13 | 22 | - 22 | |
| - 15 | - 11 | - 13 | 55 | 32 | - 35 | 11 | 11 | 15:7 |
| 58 | - 21 | 22 | - 24 | - 12 | - 41 | - 63 | 11 | |
| | | | | | | | | |

| 1 | 2 | 3 | 4 | 5 | 6 | 7 | 8 | |
|---|---|---|---|---|---|---|---|---|
| 43 | 45 | 41 | 44 | 35 | 43 | 65 | 27 | |
| 23 | - 13 | 33 | - 31 | 32 | 35 | - 21 | 61 | |
| - 11 | 32 | 12 | 22 | - 23 | - 12 | 32 | - 13 | 15:8 |
| - 11 | - 23 | - 75 | 33 | 51 | - 53 | - 23 | - 32 | |
| | | | | | | | | |

TIME: _____min _____sec    Accuracy _____/16

| | 1 | 2 | 3 | 4 | 5 | 6 | 7 | 8 | |
|---|---|---|---|---|---|---|---|---|---|
| | 44 | 55 | 75 | 24 | 56 | 55 | 75 | 63 | |
| | 53 | - 13 | - 21 | 35 | - 11 | 11 | - 31 | 12 | 15:9 |
| | - 62 | 25 | - 30 | - 10 | 32 | 33 | 53 | - 21 | © SAI Speed Math Academy, USA |
| | - 01 | - 36 | - 02 | - 24 | - 53 | - 91 | - 72 | - 32 | |
| | | | | | | | | | |

| | 1 | 2 | 3 | 4 | 5 | 6 | 7 | 8 | |
|---|---|---|---|---|---|---|---|---|---|
| | 65 | 77 | 44 | 25 | 78 | 44 | 69 | 22 | |
| | 31 | - 32 | 13 | 23 | - 52 | 22 | - 21 | 32 | 15:10 |
| | - 82 | 51 | - 05 | 11 | 31 | - 53 | 10 | 23 | © SAI Speed Math Academy, USA |
| | 31 | - 63 | - 32 | - 31 | - 53 | 35 | - 35 | - 36 | |
| | | | | | | | | | |

## Word Search – Classroom

```
K R I A H C S A I S P L C D
M E O P R S G L O B E A O O
R P A A I M H T A M E I M O
K A B E H A V E A M D C P G
S P R S A R R U L E R O U E
E O U P O T T Y D U T S T T
D O A D L A N G U A G E E I
I M M A E S C I E N C E R R
F R I E N D H L I C N E P W
S N A R T E A C H E R G O O
```

| Desk | Chair | Teacher |
|---|---|---|
| Math | Science | Study |
| Map | Globe | Pencil |
| Paper | Write | Social |
| Behave | Good | Friend |
| Smart | Ruler | Read |
| Poem | Computer | |
| | Language | |

# WEEK 15 – MIND MATH PRACTICE

Visualize

## DAY 1 – MONDAY  Accuracy _____/10 ☆

| 1 | 2 | 3 | 4 | 5 | 6 | 7 | 8 | 9 | 10 | |
|---|---|---|---|---|---|---|---|---|---|---|
| 69 | 55 | 34 | 66 | 54 | 52 | 43 | 46 | 69 | 33 | |
| - 15 | 11 | 11 | - 13 | 12 | 23 | 03 | 11 | - 25 | 21 | |
| 22 | 22 | 30 | 32 | - 11 | - 30 | - 21 | - 25 | 30 | 21 | 15:11 |
| - 30 | - 73 | - 15 | - 71 | - 23 | 53 | - 13 | 32 | - 61 | - 05 | |
|  |  |  |  |  |  |  |  |  |  | |

## DAY 2 – TUESDAY  Accuracy _____/10 ☆

| 1 | 2 | 3 | 4 | 5 | 6 | 7 | 8 | 9 | 10 | |
|---|---|---|---|---|---|---|---|---|---|---|
| 34 | 60 | 55 | 45 | 85 | 64 | 46 | 43 | 55 | 85 | |
| 13 | 13 | 11 | 21 | 11 | 30 | 51 | 23 | - 21 | - 73 | |
| 32 | 10 | - 32 | - 30 | - 63 | - 61 | - 25 | - 21 | 15 | 61 | 15:12 |
| - 54 | - 62 | - 34 | 20 | 06 | 20 | - 12 | - 13 | - 46 | 13 | |
|  |  |  |  |  |  |  |  |  |  | |

## DAY 3 – WEDNESDAY  Accuracy _____/10 ☆

| 1 | 2 | 3 | 4 | 5 | 6 | 7 | 8 | 9 | 10 | |
|---|---|---|---|---|---|---|---|---|---|---|
| 33 | 46 | 54 | 56 | 43 | 24 | 95 | 64 | 14 | 57 | |
| 25 | 21 | 33 | - 22 | 20 | 31 | - 61 | 12 | 12 | - 15 | |
| - 31 | - 55 | - 55 | - 13 | 25 | - 12 | 11 | - 53 | 13 | - 32 | 15:13 |
| 22 | 23 | 31 | 35 | - 11 | 51 | - 45 | - 21 | - 08 | 56 | |
|  |  |  |  |  |  |  |  |  |  | |

## DAY 4 – THURSDAY  Accuracy _____/10 ☆

| 1 | 2 | 3 | 4 | 5 | 6 | 7 | 8 | 9 | 10 | |
|---|---|---|---|---|---|---|---|---|---|---|
| 36 | 57 | 51 | 75 | 94 | 54 | 45 | 23 | 43 | 34 | |
| - 22 | - 13 | 16 | - 32 | 03 | 11 | 24 | 31 | 31 | 21 | |
| 70 | 30 | - 31 | 11 | - 42 | - 22 | - 38 | - 13 | - 20 | - 32 | 15:14 |
| - 11 | - 53 | - 12 | - 34 | - 11 | 35 | 22 | 36 | - 33 | 15 | |
|  |  |  |  |  |  |  |  |  |  | |

| 1 | 2 | 3 | 4 | 5 | 6 | 7 | 8 | 9 | 10 | |
|---|---|---|---|---|---|---|---|---|---|---|
| 35 | 24 | 77 | 52 | 85 | 65 | 34 | 34 | 43 | 24 | |
| - 23 | 22 | - 23 | 13 | - 61 | 31 | 22 | 30 | 21 | 21 | |
| 65 | - 23 | 21 | - 21 | 35 | - 43 | - 13 | 31 | 01 | - 30 | 15:15 |
| - 30 | 10 | - 35 | 11 | - 12 | - 11 | 31 | - 55 | - 50 | 02 | |
| | | | | | | | | | | |

| 1 | 2 | 3 | 4 | 5 | 6 | 7 | 8 | 9 | 10 | |
|---|---|---|---|---|---|---|---|---|---|---|
| 67 | 56 | 88 | 75 | 43 | 34 | 46 | 63 | 44 | 69 | |
| - 21 | 21 | - 31 | - 23 | 12 | 25 | 11 | 23 | 33 | - 12 | |
| 32 | - 31 | - 23 | 41 | - 23 | - 12 | - 23 | - 71 | - 22 | - 23 | 15:16 |
| - 35 | 13 | 52 | - 62 | 32 | - 17 | 15 | 54 | - 12 | 15 | |
| | | | | | | | | | | |

## SHAPE PUZZLE

Find out what each shape is worth.   Color the shapes.

Suggestion:  Take as many numbers of beans, blocks or coins as the right side of the equation shows and share them on the shapes to the left side of the equation.

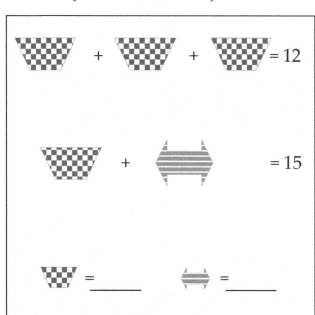

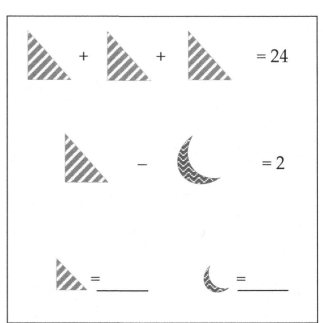

## WEEK 15 – DICTATION

**DO 6 PROBLEMS A DAY** and write answers below.

Students: Every day use abacus to calculate three dictation problems and try mind math for three dictation problems.

Teachers: Dictate any problem from this week's homework.

| 1 | 2 | 3 | 4 | 5 | 6 | 7 | 8 | 9 | 10 |
|---|---|---|---|---|---|---|---|---|---|
|   |   |   |   |   |   |   |   |   |   |

| 11 | 12 | 13 | 14 | 15 | 16 | 17 | 18 | 19 | 20 |
|---|---|---|---|---|---|---|---|---|---|
|   |   |   |   |   |   |   |   |   |   |

| 21 | 22 | 23 | 24 | 25 | 26 | 27 | 28 | 29 | 30 |
|---|---|---|---|---|---|---|---|---|---|
|   |   |   |   |   |   |   |   |   |   |

## WEEK 15 – SKILL BUILDING

Visualize the numbers on the beam in your mind and draw it to represent the numbers given.

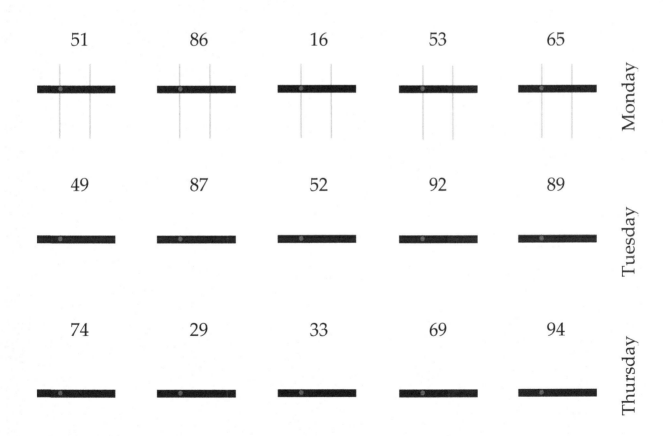

| 51 | 86 | 16 | 53 | 65 | Monday |
| 49 | 87 | 52 | 92 | 89 | Tuesday |
| 74 | 29 | 33 | 69 | 94 | Thursday |

# WEEK 16 – LESSON 15 – INTRODUCING +4 CONCEPT

### LESSON 15 –PRACTICE WORK

+4 = + 5 –1

**DAY 1 – MONDAY**

Accuracy _____/32 ☆

Use Abacus

| 1 | 2 | 3 | 4 | 5 | 6 | 7 | 8 |
|---|---|---|---|---|---|---|---|
| 01 | 02 | 03 | 04 | 05 | 12 | 44 | 31 |
| 04 | 04 | 04 | 04 | 04 | 44 | 11 | 44 |
| 10 | 20 | 30 | 40 | 50 | 11 | - 22 | - 21 |
| 40 | 40 | 40 | 40 | 40 | 12 | 13 | 34 |
|   |   |   |   |   |   |   |   |

16:1

| 1 | 2 | 3 | 4 | 5 | 6 | 7 | 8 |
|---|---|---|---|---|---|---|---|
| 41 | 23 | 13 | 13 | 22 | 31 | 53 | 46 |
| 13 | 44 | 44 | 14 | 42 | 44 | - 22 | 41 |
| - 52 | - 51 | - 21 | 31 | 24 | 13 | 43 | - 15 |
| 44 | 40 | 40 | - 56 | - 83 | - 76 | - 72 | - 30 |
|   |   |   |   |   |   |   |   |

16:2

| 1 | 2 | 3 | 4 | 5 | 6 | 7 | 8 |
|---|---|---|---|---|---|---|---|
| 25 | 21 | 11 | 34 | 11 | 23 | 21 | 41 |
| 12 | 34 | 40 | - 21 | 43 | 33 | 44 | 14 |
| 31 | 31 | 11 | 41 | 31 | - 21 | - 15 | - 21 |
| - 27 | - 55 | - 62 | - 51 | 13 | - 31 | 33 | 12 |
|   |   |   |   |   |   |   |   |

16:3

| 1 | 2 | 3 | 4 | 5 | 6 | 7 | 8 |
|---|---|---|---|---|---|---|---|
| 42 | 52 | 14 | 21 | 14 | 42 | 96 | 13 |
| 43 | 43 | 44 | 23 | 41 | 12 | - 53 | 43 |
| - 61 | - 21 | 20 | 41 | 43 | 14 | 44 | 30 |
| 13 | - 14 | - 52 | - 32 | - 95 | - 12 | - 77 | - 75 |
|   |   |   |   |   |   |   |   |

16:4

www.abacus-math.com

## DAY 2 – TUESDAY

TIME: _____ min _____ sec   Accuracy _____ /16 ☆

| 1 | 2 | 3 | 4 | 5 | 6 | 7 | 8 |
|---|---|---|---|---|---|---|---|
| 45 | 81 | 44 | 21 | 34 | 56 | 24 | 11 |
| - 13 | 11 | 41 | 42 | 31 | - 22 | 13 | 23 |
| 31 | - 50 | - 31 | - 23 | - 52 | 44 | 31 | 44 |
| - 22 | 44 | 23 | 15 | 44 | - 21 | - 17 | - 52 |
|  |  |  |  |  |  |  |  |

16:5

| 1 | 2 | 3 | 4 | 5 | 6 | 7 | 8 |
|---|---|---|---|---|---|---|---|
| 53 | 33 | 52 | 75 | 11 | 43 | 66 | 68 |
| 34 | 21 | 21 | 14 | 04 | - 12 | 30 | - 32 |
| - 13 | - 32 | 13 | - 55 | 11 | 14 | - 42 | 41 |
| - 71 | 06 | - 72 | 41 | - 26 | - 45 | 04 | - 63 |
|  |  |  |  |  |  |  |  |

16:6

## DAY 3 – WEDNESDAY

TIME: _____ min _____ sec   Accuracy _____ /16 ☆

| 1 | 2 | 3 | 4 | 5 | 6 | 7 | 8 |
|---|---|---|---|---|---|---|---|
| 33 | 52 | 62 | 13 | 98 | 22 | 56 | 34 |
| 41 | - 21 | 24 | 43 | - 32 | 34 | - 12 | 32 |
| - 30 | 34 | - 31 | - 12 | - 21 | - 13 | - 13 | - 23 |
| 14 | - 51 | - 32 | 44 | - 13 | 44 | 44 | 34 |
|  |  |  |  |  |  |  |  |

16:7

| 1 | 2 | 3 | 4 | 5 | 6 | 7 | 8 |
|---|---|---|---|---|---|---|---|
| 41 | 34 | 31 | 12 | 54 | 24 | 25 | 75 |
| 43 | 34 | 34 | 44 | 41 | 34 | 34 | - 31 |
| 11 | - 57 | - 12 | - 51 | - 92 | - 16 | - 12 | 44 |
| - 22 | 41 | - 13 | 33 | 54 | 31 | - 41 | - 36 |
|  |  |  |  |  |  |  |  |

16:8

## DAY 4 – THURSDAY

TIME: _____ min _____ sec    Accuracy _____ /16    ☆

| 1 | 2 | 3 | 4 | 5 | 6 | 7 | 8 |
|---|---|---|---|---|---|---|---|
| 32 | 31 | 15 | 82 | 42 | 31 | 23 | 95 |
| 43 | 41 | 23 | 04 | 11 | 32 | 43 | 04 |
| - 32 | - 21 | 40 | - 35 | - 21 | 33 | 32 | - 68 |
| 43 | 18 | - 31 | 04 | 32 | - 75 | - 62 | 44 |
|  |  |  |  |  |  |  |  |

16:9

| 1 | 2 | 3 | 4 | 5 | 6 | 7 | 8 |
|---|---|---|---|---|---|---|---|
| 23 | 21 | 47 | 75 | 14 | 44 | 68 | 22 |
| 43 | 54 | - 43 | - 31 | 41 | 44 | - 23 | 34 |
| - 33 | 23 | 34 | 41 | - 33 | - 32 | 42 | - 10 |
| 41 | - 73 | - 36 | 14 | 14 | - 23 | - 13 | 42 |
|  |  |  |  |  |  |  |  |

16:10

## DAY 5 – FRIDAY

TIME: _____ min _____ sec    Accuracy _____ /16    ☆

| 1 | 2 | 3 | 4 | 5 | 6 | 7 | 8 |
|---|---|---|---|---|---|---|---|
| 24 | 55 | 65 | 21 | 35 | 67 | 95 | 53 |
| 24 | - 11 | - 21 | 34 | 41 | - 11 | - 61 | 24 |
| - 08 | 24 | 14 | - 15 | - 53 | - 33 | 42 | 22 |
| 44 | - 37 | - 22 | 49 | 41 | 45 | - 53 | - 19 |
|  |  |  |  |  |  |  |  |

16:11

| 1 | 2 | 3 | 4 | 5 | 6 | 7 | 8 |
|---|---|---|---|---|---|---|---|
| 35 | 26 | 41 | 51 | 85 | 66 | 69 | 52 |
| 44 | 42 | 24 | 44 | - 63 | - 12 | - 31 | 44 |
| - 29 | - 56 | - 35 | - 72 | 54 | - 53 | - 33 | - 63 |
| - 30 | 42 | - 30 | 46 | - 35 | 74 | 94 | 54 |
|  |  |  |  |  |  |  |  |

16:12

www.abacus-math.com

## LESSON 15 – MIND MATH PRACTICE

Visualize

### DAY 1 – MONDAY

Accuracy _____/10

| 1 | 2 | 3 | 4 | 5 | 6 | 7 | 8 | 9 | 10 |
|---|---|---|---|---|---|---|---|---|---|
| 11 | 04 | 30 | 23 | 34 | 43 | 52 | 51 | 22 | 45 |
| 04 | 05 | 40 | - 10 | - 02 | - 13 | 32 | 04 | 14 | 40 |
| - 10 | - 03 | - 20 | 40 | 04 | 41 | - 54 | - 21 | - 20 | - 50 |
| 40 | - 02 | - 30 | - 20 | - 02 | - 11 | - 20 | 40 | 42 | - 01 |
|  |  |  |  |  |  |  |  |  |  |

16:13

### DAY 2 – TUESDAY

Accuracy _____/10

| 1 | 2 | 3 | 4 | 5 | 6 | 7 | 8 | 9 | 10 |
|---|---|---|---|---|---|---|---|---|---|
| 11 | 24 | 15 | 72 | 35 | 52 | 10 | 11 | 22 | 35 |
| 45 | - 11 | - 01 | - 20 | 41 | 04 | 40 | 41 | 04 | 40 |
| 10 | 40 | 42 | 44 | - 50 | 41 | 40 | - 30 | 02 | - 13 |
| - 60 | - 10 | - 03 | - 40 | - 15 | - 05 | - 70 | 14 | - 16 | - 50 |
|  |  |  |  |  |  |  |  |  |  |

16:14

### DAY 3 – WEDNESDAY

Accuracy _____/10

| 1 | 2 | 3 | 4 | 5 | 6 | 7 | 8 | 9 | 10 |
|---|---|---|---|---|---|---|---|---|---|
| 24 | 15 | 33 | 12 | 31 | 24 | 32 | 78 | 99 | 77 |
| 04 | 40 | - 21 | 44 | 14 | 40 | 24 | - 35 | - 44 | - 32 |
| - 03 | 40 | 44 | 11 | 40 | 04 | - 10 | 04 | 11 | 40 |
| - 01 | - 20 | 20 | - 10 | - 21 | 11 | 41 | - 40 | - 33 | 01 |
|  |  |  |  |  |  |  |  |  |  |

16:15

### DAY 4 – THURSDAY

Accuracy _____/10

| 1 | 2 | 3 | 4 | 5 | 6 | 7 | 8 | 9 | 10 |
|---|---|---|---|---|---|---|---|---|---|
| 15 | 39 | 10 | 22 | 13 | 43 | 47 | 23 | 31 | 14 |
| - 01 | - 03 | 44 | 42 | 41 | 40 | - 35 | 40 | 34 | 24 |
| 04 | 02 | - 02 | - 13 | 11 | - 50 | 44 | - 51 | - 10 | - 15 |
| 40 | 40 | 04 | 04 | - 53 | 04 | 10 | 24 | 24 | 40 |
|  |  |  |  |  |  |  |  |  |  |

16:16

Accuracy _____/20 ☆

| 1 | 2 | 3 | 4 | 5 | 6 | 7 | 8 | 9 | 10 |
|---|---|---|---|---|---|---|---|---|---|
| 34 | 99 | 21 | 45 | 12 | 53 | 11 | 31 | 63 | 32 |
| 14 | - 33 | 24 | - 21 | 40 | 04 | 34 | 40 | 04 | 44 |
| - 32 | - 11 | - 23 | 40 | - 30 | - 10 | 20 | - 21 | - 15 | - 13 |
| - 15 | 44 | 04 | - 30 | 40 | - 20 | 10 | - 30 | 04 | 14 |
|  |  |  |  |  |  |  |  |  |  |

16:17  © SAI Speed Math Academy, USA

| 1 | 2 | 3 | 4 | 5 | 6 | 7 | 8 | 9 | 10 |
|---|---|---|---|---|---|---|---|---|---|
| 55 | 24 | 20 | 21 | 46 | 29 | 41 | 23 | 25 | 54 |
| 12 | 40 | 44 | 40 | - 22 | 40 | - 30 | 40 | - 11 | 24 |
| - 51 | - 01 | - 51 | - 10 | 40 | - 14 | 44 | - 51 | 44 | 10 |
| 22 | 14 | 04 | 34 | - 10 | - 55 | 02 | 40 | - 30 | - 55 |
|  |  |  |  |  |  |  |  |  |  |

16:18  © SAI Speed Math Academy, USA

## SHAPE PUZZLE

Find out what each shape is worth.   Color the shapes.

Suggestion: Take as many numbers of beans, blocks or coins as the right side of the equation shows and share them on the shapes to the left side of the equation.

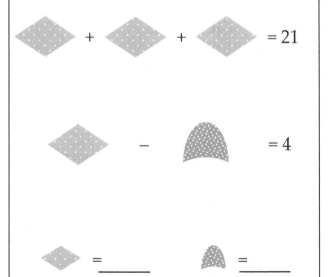

www.abacus-math.com

## LESSON 15 – DICTATION

**DO 6 PROBLEMS A DAY** and write answers below.

Students: This week try calculating all dictation problems in mind. If problem is challenging, do it on abacus and study the beads movement. Then try them in mind.

Teachers: Dictate any problem from this week's homework.

| 1 | 2 | 3 | 4 | 5 | 6 | 7 | 8 | 9 | 10 |
|---|---|---|---|---|---|---|---|---|---|
|   |   |   |   |   |   |   |   |   |   |

| 11 | 12 | 13 | 14 | 15 | 16 | 17 | 18 | 19 | 20 |
|----|----|----|----|----|----|----|----|----|----|
|    |    |    |    |    |    |    |    |    |    |

| 21 | 22 | 23 | 24 | 25 | 26 | 27 | 28 | 29 | 30 |
|----|----|----|----|----|----|----|----|----|----|
|    |    |    |    |    |    |    |    |    |    |

## LESSON 15 – SKILL BUILDING

Visualize the numbers on the beam in your mind and draw it to represent the numbers given.

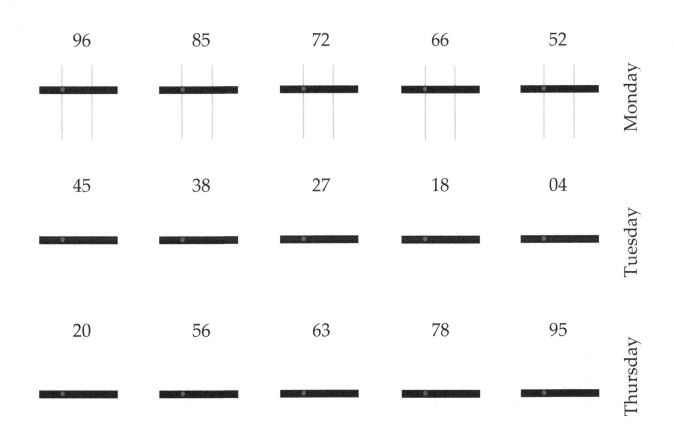

Monday: 96   85   72   66   52

Tuesday: 45   38   27   18   04

Thursday: 20   56   63   78   95

## LESSON 16 – PRACTICE WORK

$$-4 = -5 + 1$$

**DAY 1 – MONDAY**　　　Accuracy _____ /32　☆　　Use Abacus

| 1 | 2 | 3 | 4 | 5 | 6 | 7 | 8 |
|---|---|---|---|---|---|---|---|
| 02 | 03 | 68 | 74 | 55 | 15 | 99 | 88 |
| 02 | 02 | - 31 | - 34 | - 43 | 31 | - 44 | - 33 |
| - 04 | 01 | - 11 | 16 | 52 | - 44 | - 44 | - 44 |
| 17 | - 04 | - 04 | - 41 | - 41 | 12 | 33 | 22 |
|  |  |  |  |  |  |  |  |

17:1

| 1 | 2 | 3 | 4 | 5 | 6 | 7 | 8 |
|---|---|---|---|---|---|---|---|
| 66 | 51 | 47 | 42 | 65 | 48 | 54 | 53 |
| - 44 | - 10 | - 34 | 12 | - 04 | - 44 | 21 | 32 |
| 22 | 05 | 55 | - 41 | 11 | 61 | - 44 | - 44 |
| - 14 | - 44 | - 43 | 32 | - 42 | - 43 | - 20 | 33 |
|  |  |  |  |  |  |  |  |

17:2

| 1 | 2 | 3 | 4 | 5 | 6 | 7 | 8 |
|---|---|---|---|---|---|---|---|
| 45 | 15 | 14 | 47 | 14 | 75 | 69 | 13 |
| 31 | 41 | 31 | 12 | 64 | - 41 | - 14 | 53 |
| - 43 | - 14 | 41 | - 14 | - 14 | - 31 | 21 | - 24 |
| - 13 | - 32 | - 25 | - 14 | 31 | 21 | - 43 | - 12 |
|  |  |  |  |  |  |  |  |

17:3

| 1 | 2 | 3 | 4 | 5 | 6 | 7 | 8 |
|---|---|---|---|---|---|---|---|
| 41 | 87 | 53 | 58 | 71 | 56 | 44 | 13 |
| 12 | - 04 | - 42 | - 44 | 05 | - 42 | 32 | 41 |
| - 41 | - 41 | 71 | 12 | - 34 | 53 | - 34 | 12 |
| - 12 | - 31 | - 40 | - 14 | - 21 | - 47 | - 42 | - 24 |
|  |  |  |  |  |  |  |  |

17:4

## DAY 2 – TUESDAY

TIME: _____ min _____ sec    Accuracy _____/16 ☆

| 1 | 2 | 3 | 4 | 5 | 6 | 7 | 8 | |
|---|---|---|---|---|---|---|---|---|
| 16 | 55 | 14 | 42 | 84 | 43 | 25 | 78 | |
| - 04 | - 11 | 11 | 42 | - 20 | 21 | - 12 | 11 | 17:5 |
| 22 | 22 | 31 | - 31 | - 43 | - 11 | 32 | - 24 | |
| 41 | - 33 | - 42 | - 41 | 16 | - 40 | - 44 | - 31 | |
| | | | | | | | | |

| 1 | 2 | 3 | 4 | 5 | 6 | 7 | 8 | |
|---|---|---|---|---|---|---|---|---|
| 26 | 14 | 24 | 43 | 33 | 47 | 73 | 78 | |
| 43 | 12 | 32 | 11 | 42 | - 24 | - 31 | - 42 | 17:6 |
| - 54 | 11 | - 41 | 35 | - 23 | - 13 | 15 | 30 | |
| - 11 | 12 | - 14 | - 44 | - 21 | 71 | - 34 | - 64 | |
| | | | | | | | | |

## DAY 3 – WEDNESDAY

TIME: _____ min _____ sec    Accuracy _____/16 ☆

| 1 | 2 | 3 | 4 | 5 | 6 | 7 | 8 | |
|---|---|---|---|---|---|---|---|---|
| 55 | 87 | 22 | 43 | 46 | 63 | 25 | 15 | |
| - 41 | - 12 | 12 | 24 | 13 | - 22 | - 14 | 43 | 17:7 |
| 15 | - 14 | 15 | - 31 | - 32 | 46 | 82 | - 34 | |
| - 23 | 25 | - 39 | - 34 | - 12 | - 13 | - 43 | 22 | |
| | | | | | | | | |

| 1 | 2 | 3 | 4 | 5 | 6 | 7 | 8 | |
|---|---|---|---|---|---|---|---|---|
| 57 | 34 | 95 | 34 | 99 | 31 | 88 | 85 | |
| - 16 | 23 | - 12 | 53 | - 54 | 44 | - 42 | - 54 | 17:8 |
| 11 | - 43 | - 21 | - 45 | 44 | - 32 | - 31 | 12 | |
| - 20 | - 11 | - 41 | - 41 | - 13 | 14 | - 14 | 46 | |
| | | | | | | | | |

## DAY 4 – THURSDAY

TIME: _____ min _____ sec   Accuracy _____ /16

| 1 | 2 | 3 | 4 | 5 | 6 | 7 | 8 |
|---|---|---|---|---|---|---|---|
| 77 | 54 | 43 | 32 | 46 | 76 | 55 | 52 |
| - 14 | 13 | 32 | 34 | 11 | - 41 | 31 | 33 |
| 36 | - 41 | - 41 | - 41 | - 24 | 12 | - 52 | - 41 |
| - 21 | - 24 | 35 | - 12 | - 12 | 50 | 43 | 12 |
|  |  |  |  |  |  |  |  |

17:9 © SAI Speed Math Academy, USA

| 1 | 2 | 3 | 4 | 5 | 6 | 7 | 8 |
|---|---|---|---|---|---|---|---|
| 85 | 92 | 68 | 75 | 84 | 44 | 66 | 69 |
| - 43 | 04 | - 44 | - 44 | 15 | 31 | - 43 | 30 |
| 12 | - 43 | 31 | 11 | - 24 | - 42 | 32 | - 24 |
| - 13 | - 40 | - 53 | 17 | - 41 | 55 | - 21 | - 44 |
|  |  |  |  |  |  |  |  |

17:10 © SAI Speed Math Academy, USA

## DAY 5 – FRIDAY

TIME: _____ min _____ sec   Accuracy _____ /16

| 1 | 2 | 3 | 4 | 5 | 6 | 7 | 8 |
|---|---|---|---|---|---|---|---|
| 64 | 26 | 76 | 21 | 45 | 53 | 75 | 56 |
| 13 | - 14 | - 24 | 35 | - 11 | 11 | - 24 | 12 |
| - 42 | 53 | - 51 | - 46 | 43 | - 44 | 17 | - 43 |
| - 14 | - 41 | 94 | 57 | - 64 | - 20 | - 64 | - 04 |
|  |  |  |  |  |  |  |  |

17:11 © SAI Speed Math Academy, USA

| 1 | 2 | 3 | 4 | 5 | 6 | 7 | 8 |
|---|---|---|---|---|---|---|---|
| 15 | 46 | 41 | 14 | 75 | 17 | 67 | 22 |
| 31 | 12 | 16 | 73 | - 54 | 42 | - 44 | 64 |
| - 24 | - 45 | - 24 | - 11 | 35 | - 16 | 35 | - 14 |
| 36 | 32 | - 31 | - 24 | - 14 | 34 | - 48 | - 42 |
|  |  |  |  |  |  |  |  |

17:12 © SAI Speed Math Academy, USA

## LESSON 16 – MIND MATH PRACTICE

Visualize

### DAY 1 – MONDAY  Accuracy _____/10 ☆

| 1 | 2 | 3 | 4 | 5 | 6 | 7 | 8 | 9 | 10 | |
|---|---|---|---|---|---|---|---|---|---|---|
| 53 | 54 | 43 | 23 | 32 | 11 | 24 | 55 | 75 | 35 | |
| 10 | 05 | 31 | - 10 | - 21 | 35 | - 11 | - 04 | - 20 | 21 | 17:13 |
| - 40 | - 30 | - 20 | 21 | 11 | 10 | 32 | 32 | 10 | - 40 | |
| 10 | - 20 | - 40 | - 24 | - 22 | - 40 | - 14 | - 33 | - 44 | - 12 | |
| | | | | | | | | | | |

### DAY 2 – TUESDAY  Accuracy _____/10 ☆

| 1 | 2 | 3 | 4 | 5 | 6 | 7 | 8 | 9 | 10 | |
|---|---|---|---|---|---|---|---|---|---|---|
| 63 | 52 | 54 | 22 | 45 | 34 | 99 | 24 | 45 | 32 | |
| - 13 | 32 | 31 | 13 | 23 | 13 | - 33 | 21 | - 21 | 35 | 17:14 |
| 01 | - 40 | - 20 | - 34 | - 40 | - 32 | - 44 | - 24 | 30 | - 40 | |
| - 41 | - 40 | - 40 | 32 | - 06 | - 14 | 31 | 01 | - 40 | - 07 | |
| | | | | | | | | | | |

### DAY 3 – WEDNESDAY  Accuracy _____/10 ☆

| 1 | 2 | 3 | 4 | 5 | 6 | 7 | 8 | 9 | 10 | |
|---|---|---|---|---|---|---|---|---|---|---|
| 44 | 55 | 35 | 12 | 46 | 57 | 33 | 64 | 26 | 77 | |
| 30 | - 40 | - 04 | 30 | 31 | - 20 | 23 | 31 | 21 | - 32 | 17:15 |
| - 10 | 10 | 11 | 51 | - 40 | - 04 | - 40 | - 41 | - 04 | - 04 | |
| - 40 | - 23 | 11 | - 40 | 10 | 15 | 13 | - 40 | - 02 | 03 | |
| | | | | | | | | | | |

### DAY 4 – THURSDAY  Accuracy _____/10 ☆

| 1 | 2 | 3 | 4 | 5 | 6 | 7 | 8 | 9 | 10 | |
|---|---|---|---|---|---|---|---|---|---|---|
| 15 | 59 | 44 | 22 | 73 | 43 | 47 | 23 | 34 | 14 | |
| - 04 | - 30 | 12 | 32 | - 30 | 21 | - 34 | 31 | 01 | 22 | 17:16 |
| 32 | 20 | - 40 | - 14 | 11 | - 14 | 05 | - 41 | - 14 | - 14 | |
| - 40 | - 34 | - 04 | 30 | - 40 | - 40 | - 04 | 20 | 26 | 11 | |
| | | | | | | | | | | |

Accuracy _____ /20 ☆

| 1 | 2 | 3 | 4 | 5 | 6 | 7 | 8 | 9 | 10 |
|---|---|---|---|---|---|---|---|---|---|
| 54 | 15 | 11 | 22 | 35 | 53 | 11 | 31 | 15 | 32 |
| - 13 | 30 | 21 | 22 | 40 | 21 | 34 | 40 | - 11 | 15 |
| 25 | 54 | - 31 | 33 | - 13 | - 13 | 20 | - 21 | 53 | - 43 |
| - 20 | - 55 | 13 | - 44 | - 40 | - 21 | 01 | - 30 | - 40 | 12 |
|  |  |  |  |  |  |  |  |  |  |

17:17

| 1 | 2 | 3 | 4 | 5 | 6 | 7 | 8 | 9 | 10 |
|---|---|---|---|---|---|---|---|---|---|
| 47 | 24 | 44 | 85 | 46 | 34 | 41 | 40 | 12 | 51 |
| 10 | 30 | 20 | - 12 | - 22 | 12 | - 30 | 22 | 62 | 21 |
| - 31 | - 40 | - 51 | - 40 | 01 | - 24 | 23 | - 41 | - 40 | 12 |
| - 04 | 30 | 30 | - 22 | - 04 | 01 | - 04 | 20 | - 11 | - 44 |
|  |  |  |  |  |  |  |  |  |  |

17:18

## SHAPE PUZZLE

Find out what each shape is worth.   Color the shapes.

Suggestion:  Take as many numbers of beans, blocks or coins as the right side of the equation shows and share them on the shapes to the left side of the equation.

www.abacus-math.com

## LESSON 16 – DICTATION

**DO 6 PROBLEMS A DAY** and write answers below.

Students: This week try calculating all dictation problems in mind. If problem is challenging, do it on abacus and study the beads movement. Then try them in mind.

Teachers: Dictate any problem from this week's homework.

| 1 | 2 | 3 | 4 | 5 | 6 | 7 | 8 | 9 | 10 |
|---|---|---|---|---|---|---|---|---|----|
|   |   |   |   |   |   |   |   |   |    |

| 11 | 12 | 13 | 14 | 15 | 16 | 17 | 18 | 19 | 20 |
|----|----|----|----|----|----|----|----|----|----|
|    |    |    |    |    |    |    |    |    |    |

| 21 | 22 | 23 | 24 | 25 | 26 | 27 | 28 | 29 | 30 |
|----|----|----|----|----|----|----|----|----|----|
|    |    |    |    |    |    |    |    |    |    |

## LESSON 16 – SKILL BUILDING

Visualize the numbers on the beam in your mind and draw it to represent the numbers given.

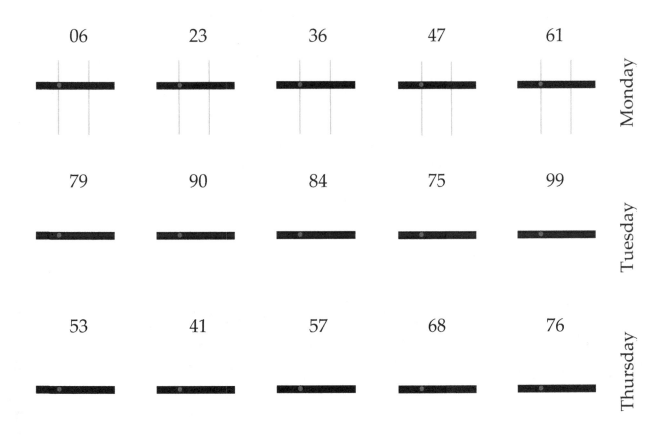

| 06 | 23 | 36 | 47 | 61 | Monday |

| 79 | 90 | 84 | 75 | 99 | Tuesday |

| 53 | 41 | 57 | 68 | 76 | Thursday |

# WEEK 18 – SKILL BUILDING

## WEEK 18 – PRACTICE WORK

### DAY 1 – MONDAY

TIME: _____ min _____ sec    Accuracy _____ /16

| 1 | 2 | 3 | 4 | 5 | 6 | 7 | 8 |
|---|---|---|---|---|---|---|---|
| 56 | 26 | 16 | 43 | 66 | 75 | 23 | 65 |
| - 42 | - 14 | 41 | 43 | - 42 | - 44 | 34 | - 34 |
| - 13 | 42 | - 24 | - 22 | 31 | - 21 | 41 | 24 |
| 34 | - 31 | - 31 | 11 | - 12 | 14 | - 75 | - 45 |
|  |  |  |  |  |  |  |  |

*18:1*

| 1 | 2 | 3 | 4 | 5 | 6 | 7 | 8 |
|---|---|---|---|---|---|---|---|
|  |  | 24 | 49 | 73 | 46 | 55 | 49 |
| 33 | 21 | 24 | - 31 | 14 | - 14 | - 42 | 30 |
| 14 | 34 | - 33 | 60 | - 42 | 46 | 34 | - 25 |
| - 42 | - 41 | - 12 | - 44 | - 41 | - 24 | - 43 | - 32 |
| 10 | 73 | 82 | 51 | 35 | 45 | 11 | 55 |
|  |  |  |  |  |  |  |  |

*18:2*

### DAY 2 – TUESDAY

TIME: _____ min _____ sec    Accuracy _____ /16

| 1 | 2 | 3 | 4 | 5 | 6 | 7 | 8 |
|---|---|---|---|---|---|---|---|
| 44 | 18 | 43 | 58 | 56 | 12 | 78 | 74 |
| 33 | - 04 | 14 | - 34 | - 24 | 34 | - 14 | 22 |
| - 44 | 13 | - 40 | 23 | 33 | - 11 | - 32 | - 53 |
| 22 | 41 | - 14 | - 44 | - 14 | - 24 | - 32 | 14 |
|  |  |  |  |  |  |  |  |

*18:3*

| 1 | 2 | 3 | 4 | 5 | 6 | 7 | 8 |
|---|---|---|---|---|---|---|---|
| 23 | 34 | 13 | 41 | 29 | 55 | 32 | 21 |
| 44 | 24 | 41 | 44 | - 14 | - 12 | 43 | 68 |
| 11 | - 16 | 23 | 12 | 41 | 31 | - 12 | - 35 |
| - 42 | 21 | - 32 | - 41 | - 24 | 14 | - 41 | 13 |
| - 24 | 33 | 44 | 12 | 66 | - 78 | 62 | - 45 |
|  |  |  |  |  |  |  |  |

*18:4*

## DAY 3 – WEDNESDAY

TIME: _____ min _____ sec   Accuracy _____ /16 ☆

18:5

| 1 | 2 | 3 | 4 | 5 | 6 | 7 | 8 |
|---|---|---|---|---|---|---|---|
|  |  |  |  |  |  | 62 | 13 |
| 15 | 63 | 57 | 85 | 34 | 52 | 24 | 43 |
| 13 | 34 | 41 | - 31 | 41 | - 21 | - 31 | - 14 |
| 30 | - 12 | - 34 | - 23 | - 24 | 34 | - 32 | 42 |
| - 42 | - 14 | 13 | 64 | 15 | - 24 | 55 | - 51 |
|  |  |  |  |  |  |  |  |

18:6

| 1 | 2 | 3 | 4 | 5 | 6 | 7 | 8 |
|---|---|---|---|---|---|---|---|
| 98 | 34 | 56 | 14 | 11 | 53 | 33 | 41 |
| - 32 | 31 | - 24 | 13 | 23 | 34 | 21 | 45 |
| - 21 | - 23 | 44 | 41 | 44 | - 13 | - 32 | - 63 |
| - 13 | 44 | - 21 | - 17 | - 23 | - 41 | 06 | 12 |
| 46 | - 65 | 43 | 48 | - 23 | 55 | 51 | 11 |
|  |  |  |  |  |  |  |  |

## DAY 4 – THURSDAY

TIME: _____ min _____ sec   Accuracy _____ /16 ☆

18:7

| 1 | 2 | 3 | 4 | 5 | 6 | 7 | 8 |
|---|---|---|---|---|---|---|---|
|  |  |  | 45 | 13 | 44 | 21 | 24 |
| 11 | 22 | 23 | - 13 | 43 | 41 | 44 | 31 |
| 24 | 24 | 24 | 31 | - 12 | - 31 | - 23 | - 42 |
| 11 | - 11 | - 31 | - 22 | - 13 | 25 | 15 | 15 |
| 41 | 44 | 42 | 15 | 65 | - 19 | - 41 | 51 |
|  |  |  |  |  |  |  |  |

18:8

| 1 | 2 | 3 | 4 | 5 | 6 | 7 | 8 |
|---|---|---|---|---|---|---|---|
| 65 | 25 | 35 | 44 | 13 | 23 | 32 | 59 |
| - 41 | - 13 | 44 | 11 | 44 | 41 | 44 | - 27 |
| 11 | 75 | - 55 | - 22 | - 12 | - 52 | - 51 | 64 |
| 54 | - 20 | 21 | 13 | 33 | 44 | 44 | - 23 |
| - 46 | 12 | - 33 | 13 | 21 | - 15 | - 59 | - 32 |
|  |  |  |  |  |  |  |  |

TIME: _____min _____sec   Accuracy _____/16 ☆

| 1 | 2 | 3 | 4 | 5 | 6 | 7 | 8 |
|---|---|---|---|---|---|---|---|
| 25 | 56 | 34 | 14 | 34 | 77 | 46 | 63 |
| 34 | - 14 | 32 | 44 | 55 | - 24 | 13 | - 22 |
| - 43 | - 31 | - 23 | - 55 | - 87 | - 01 | - 32 | 46 |
| - 12 | 44 | 34 | 71 | 14 | - 32 | - 12 | - 13 |
|   |   |   |   |   |   |   |   |

18:9
© SAI Speed Math Academy, USA

| 1 | 2 | 3 | 4 | 5 | 6 | 7 | 8 |
|---|---|---|---|---|---|---|---|
| 99 | 77 | 13 | 21 | 11 | 57 | 25 | 15 |
| - 65 | - 45 | 44 | 24 | 78 | - 44 | - 14 | 43 |
| 31 | 51 | 31 | 21 | - 44 | 22 | 82 | - 34 |
| - 42 | - 63 | - 05 | 11 | 31 | 31 | - 43 | 22 |
| 31 | 46 | - 32 | - 36 | - 13 | - 65 | - 10 | 43 |
|   |   |   |   |   |   |   |   |

18:10
© SAI Speed Math Academy, USA

## WEEK 18 – DICTATION

**DO 6 PROBLEMS A DAY** and write answers below.

Students: Try calculating all dictation problems in mind.

Teachers: Dictate any problem from this week's homework.

| 1 | 2 | 3 | 4 | 5 | 6 | 7 | 8 | 9 | 10 |
|---|---|---|---|---|---|---|---|---|----|
|   |   |   |   |   |   |   |   |   |    |

| 11 | 12 | 13 | 14 | 15 | 16 | 17 | 18 | 19 | 20 |
|----|----|----|----|----|----|----|----|----|----|
|    |    |    |    |    |    |    |    |    |    |

| 21 | 22 | 23 | 24 | 25 | 26 | 27 | 28 | 29 | 30 |
|----|----|----|----|----|----|----|----|----|----|
|    |    |    |    |    |    |    |    |    |    |

# WEEK 18 – MIND MATH PRACTICE

Visualize

## DAY 1 – MONDAY

Accuracy _____/10

| 1 | 2 | 3 | 4 | 5 | 6 | 7 | 8 | 9 | 10 |
|---|---|---|---|---|---|---|---|---|----|
| 54 | 42 | 34 | 34 | 64 | 45 | 33 | 52 | 24 | 42 |
| - 13 | 13 | 20 | 14 | 13 | - 21 | - 11 | 13 | 21 | 22 |
| 42 | - 11 | - 42 | - 03 | - 04 | 24 | 40 | 20 | 40 | - 41 |
| - 32 | - 11 | 23 | - 12 | - 13 | - 23 | 03 | - 11 | - 33 | - 21 |
|  |  |  |  |  |  |  |  |  |  |

18:11

## DAY 2 – TUESDAY

Accuracy _____/10

| 1 | 2 | 3 | 4 | 5 | 6 | 7 | 8 | 9 | 10 |
|---|---|---|---|---|---|---|---|---|----|
| 24 | 87 | 43 | 59 | 64 | 45 | 34 | 33 | 44 | 51 |
| 13 | - 42 | 23 | - 13 | 24 | - 04 | 30 | 21 | 21 | 25 |
| 21 | 04 | - 41 | - 32 | - 31 | 21 | 31 | - 41 | - 32 | - 31 |
| - 31 | - 42 | - 21 | - 11 | - 24 | - 12 | - 54 | - 11 | - 11 | - 44 |
|  |  |  |  |  |  |  |  |  |  |

18:12

## DAY 3 – WEDNESDAY

Accuracy _____/10

| 1 | 2 | 3 | 4 | 5 | 6 | 7 | 8 | 9 | 10 |
|---|---|---|---|---|---|---|---|---|----|
| 23 | 45 | 42 | 43 | 66 | 14 | 85 | 55 | 58 | 64 |
| 23 | - 31 | 21 | 31 | - 21 | 40 | - 32 | - 32 | - 47 | 33 |
| 42 | 14 | - 32 | - 21 | 42 | - 30 | - 31 | 50 | 16 | - 45 |
| - 15 | 40 | 15 | 34 | - 11 | - 12 | 14 | - 41 | 31 | - 41 |
|  |  |  |  |  |  |  |  |  |  |

18:13

## DAY 4 – THURSDAY

Accuracy _____/10

| 1 | 2 | 3 | 4 | 5 | 6 | 7 | 8 | 9 | 10 |
|---|---|---|---|---|---|---|---|---|----|
| 89 | 24 | 11 | 53 | 33 | 45 | 78 | 55 | 43 | 42 |
| - 33 | 23 | 23 | 34 | 21 | 11 | - 30 | - 41 | 12 | 12 |
| 40 | 41 | 44 | - 13 | - 32 | 40 | - 14 | 15 | - 14 | 02 |
| - 21 | - 87 | - 78 | - 71 | 17 | - 55 | 54 | - 23 | - 31 | - 34 |
|  |  |  |  |  |  |  |  |  |  |

18:14

| 1 | 2 | 3 | 4 | 5 | 6 | 7 | 8 | 9 | 10 | |
|---|---|---|---|---|---|---|---|---|---|---|
| 21 | 51 | 41 | 85 | 22 | 54 | 61 | 33 | 35 | 24 | |
| 45 | 14 | 34 | - 41 | 43 | 11 | 34 | 41 | - 21 | 24 | |
| - 14 | - 42 | - 41 | 13 | - 21 | 13 | - 61 | 21 | 63 | - 43 | 18:15 |
| - 40 | 21 | - 24 | - 47 | 54 | - 24 | - 14 | - 94 | - 04 | - 01 | |
| | | | | | | | | | | |

© SAI Speed Math Academy, USA

| 1 | 2 | 3 | 4 | 5 | 6 | 7 | 8 | 9 | 10 | |
|---|---|---|---|---|---|---|---|---|---|---|
| 21 | 42 | 54 | 62 | 56 | 44 | 82 | 15 | 88 | 41 | |
| 66 | 34 | 30 | - 21 | 13 | 15 | 04 | 30 | - 21 | 42 | |
| - 74 | - 65 | - 01 | 55 | - 24 | - 29 | - 70 | 23 | - 63 | - 53 | 18:16 |
| 23 | 43 | - 21 | - 40 | 11 | - 20 | - 14 | - 64 | - 04 | - 20 | |
| | | | | | | | | | | |

© SAI Speed Math Academy, USA

## SHAPE PUZZLE

Find out what each shape is worth.   Color the shapes.

Suggestion: Take as many numbers of beans, blocks or coins as the right side of the equation shows and share them on the shapes to the left side of the equation.

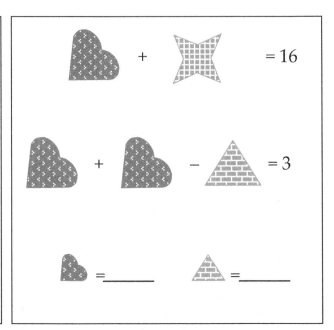

# WEEK 19 – LESSON 17 – INTRODUCING – HUNDRED'S PLACE

## PLACE VALUE PRACTICE

**What place value are 6 in the following numbers? Write the name of the place value on the given line.**

265 →  ___Tens___

603 →  _____

362 →  _____

596 →  _____

156 →  _____

649 →  _____

446 →  _____

260 →  _____

648 →  _____

106 →  _____

869 →  _____

609 →  _____

## PLACE VALUE PRACTICE

| | |
|---|---|
| 231 = _2_ hundred _3_ tens _1_ ones | 381 = ___ hundred ___ tens ___ ones |
| 340 = ___ hundred ___ tens ___ ones | 402 = ___ hundred ___ tens ___ ones |
| 472 = ___ hundred ___ tens ___ ones | 578 = ___ hundred ___ tens ___ ones |
| 055 = ___ hundred ___ tens ___ ones | 690 = ___ hundred ___ tens ___ ones |
| 615 = ___ hundred ___ tens ___ ones | 775 = ___ hundred ___ tens ___ ones |
| 708 = ___ hundred ___ tens ___ ones | 832 = ___ hundred ___ tens ___ ones |
| 850 = ___ hundred ___ tens ___ ones | 900 = ___ hundred ___ tens ___ ones |
| 916 = ___ hundred ___ tens ___ ones | 171 = ___ hundred ___ tens ___ ones |
| 105 = ___ hundred ___ tens ___ ones | 067 = ___ hundred ___ tens ___ ones |
| 268 = ___ hundred ___ tens ___ ones | 599 = ___ hundred ___ tens ___ ones |

# LESSON 17 –PRACTICE WORK

Use Abacus

## DAY 1 – MONDAY

TIME: _____min _____sec    Accuracy _____/16

| 1 | 2 | 3 | 4 | 5 | 6 | 7 | 8 | |
|---|---|---|---|---|---|---|---|---|
| 55 | 43 | 79 | 41 | 25 | 36 | 213 | 125 | |
| - 41 | 42 | - 12 | 24 | - 14 | 13 | - 112 | 423 | 19:1 |
| 15 | - 34 | - 02 | - 31 | 82 | - 32 | 436 | - 324 | |
| - 23 | - 31 | - 34 | - 34 | - 43 | - 14 | - 123 | 222 | |
| | | | | | | | | |

| 1 | 2 | 3 | 4 | 5 | 6 | 7 | 8 | |
|---|---|---|---|---|---|---|---|---|
| | | 20 | 03 | 68 | 99 | 55 | 85 | |
| 357 | 345 | 42 | 32 | - 31 | - 74 | - 43 | - 14 | |
| - 016 | 232 | 02 | 41 | - 11 | - 04 | 52 | 03 | 19:2 |
| 311 | - 433 | 14 | - 04 | - 04 | 16 | - 41 | - 30 | |
| - 020 | - 114 | - 37 | 13 | 56 | 41 | 55 | 15 | |
| | | | | | | | | |

## DAY 2 – TUESDAY

TIME: _____min _____sec    Accuracy _____/16

| 1 | 2 | 3 | 4 | 5 | 6 | 7 | 8 | |
|---|---|---|---|---|---|---|---|---|
| 71 | 56 | 44 | 75 | 66 | 43 | 161 | 887 | |
| 05 | - 42 | 32 | - 41 | - 34 | 33 | 412 | - 204 | |
| - 34 | 53 | - 34 | - 31 | 21 | - 24 | - 431 | - 241 | 19:3 |
| - 21 | - 47 | - 42 | 21 | - 41 | - 22 | - 132 | - 231 | |
| | | | | | | | | |

| 1 | 2 | 3 | 4 | 5 | 6 | 7 | 8 | |
|---|---|---|---|---|---|---|---|---|
| | | 35 | 24 | 34 | 72 | 64 | 68 | |
| 503 | 581 | - 13 | 34 | 24 | 14 | 13 | - 34 | |
| - 402 | - 441 | 24 | - 11 | - 12 | - 23 | - 14 | 15 | 19:4 |
| 791 | 125 | - 35 | - 13 | 23 | - 23 | - 31 | 40 | |
| - 430 | - 145 | 41 | 44 | - 55 | 32 | 43 | - 27 | |
| | | | | | | | | |

| 1 | 2 | 3 | 4 | 5 | 6 | 7 | 8 |
|---|---|---|---|---|---|---|---|
| 23 | 74 | 15 | 51 | 77 | 65 | 979 | 555 |
| 22 | - 34 | 31 | - 10 | - 33 | - 42 | - 414 | - 413 |
| 21 | 16 | - 44 | 05 | 11 | 20 | - 444 | 542 |
| - 34 | - 41 | 12 | - 44 | 22 | - 13 | 353 | - 481 |
|  |  |  |  |  |  |  |  |

19:5 © SAI Speed Math Academy, USA

| 1 | 2 | 3 | 4 | 5 | 6 | 7 | 8 |
|---|---|---|---|---|---|---|---|
|  |  | 14 | 85 | 43 | 34 | 45 | 514 |
| 152 | 698 | 43 | - 32 | 21 | 22 | 13 | 252 |
| 102 | - 311 | - 31 | - 31 | - 32 | - 32 | 21 | - 310 |
| - 124 | - 111 | - 12 | 14 | 13 | 21 | - 34 | - 210 |
| 117 | - 034 | 44 | 22 | - 11 | 11 | 10 | 343 |
|  |  |  |  |  |  |  |  |

19:6 © SAI Speed Math Academy, USA

| 1 | 2 | 3 | 4 | 5 | 6 | 7 | 8 |
|---|---|---|---|---|---|---|---|
| 23 | 43 | 99 | 77 | 95 | 43 | 703 | 349 |
| 34 | 11 | - 54 | - 44 | - 12 | 42 | - 301 | 530 |
| - 16 | - 13 | 44 | - 03 | - 21 | - 33 | 105 | - 457 |
| 17 | - 21 | - 13 | 11 | - 41 | - 41 | - 304 | - 412 |
|  |  |  |  |  |  |  |  |

19:7 © SAI Speed Math Academy, USA

| 1 | 2 | 3 | 4 | 5 | 6 | 7 | 8 |
|---|---|---|---|---|---|---|---|
|  |  | 45 | 61 | 33 | 35 | 245 | 127 |
| 318 | 888 | 11 | 34 | 41 | - 21 | 441 | 471 |
| 441 | - 432 | 31 | - 61 | 21 | 63 | - 233 | - 213 |
| - 327 | - 321 | - 43 | - 14 | - 61 | - 24 | - 012 | - 32 |
| 144 | - 114 | 11 | 66 | 22 | - 13 | 324 | - 140 |
|  |  |  |  |  |  |  |  |

19:8 © SAI Speed Math Academy, USA

## DAY 5 – FRIDAY

TIME: _____min _____sec    Accuracy _____/16    ⭐

| 1 | 2 | 3 | 4 | 5 | 6 | 7 | 8 | |
|---|---|---|---|---|---|---|---|---|
| 44 | 45 | 88 | 34 | 66 | 45 | 34 | 33 | |
| - 13 | - 31 | - 11 | 32 | - 21 | - 12 | 31 | 35 | |
| 42 | 40 | - 44 | - 21 | 40 | 21 | 31 | - 41 | 19:9 |
| - 51 | - 13 | 22 | 14 | - 34 | - 33 | - 54 | - 13 | |
| - 22 | 25 | - 55 | - 55 | - 41 | 04 | 11 | 05 | |
| | | | | | | | | |

| 1 | 2 | 3 | 4 | 5 | 6 | 7 | 8 | |
|---|---|---|---|---|---|---|---|---|
| | | 32 | 61 | 21 | 25 | 169 | 262 | |
| 625 | 757 | 44 | 25 | 78 | 44 | - 21 | 302 | |
| 301 | - 312 | 13 | - 41 | - 52 | 20 | 210 | 213 | 19:10 |
| - 802 | 501 | - 45 | 10 | 40 | - 75 | - 35 | - 336 | |
| 351 | - 643 | 11 | - 31 | - 64 | 54 | 56 | - 441 | |
| | | | | | | | | |

## LESSON 17 – DICTATION

**DO 6 PROBLEMS A DAY** and write answers below.

Students: Try calculating all dictation problems in mind.

Teachers: Dictate any two digit problem from this week's homework.

| 1 | 2 | 3 | 4 | 5 | 6 | 7 | 8 | 9 | 10 |
|---|---|---|---|---|---|---|---|---|----|
| | | | | | | | | | |

| 11 | 12 | 13 | 14 | 15 | 16 | 17 | 18 | 19 | 20 |
|----|----|----|----|----|----|----|----|----|----|
| | | | | | | | | | |

| 21 | 22 | 23 | 24 | 25 | 26 | 27 | 28 | 29 | 30 |
|----|----|----|----|----|----|----|----|----|----|
| | | | | | | | | | |

# LESSON 17 – MIND MATH PRACTICE

## DAY 1 – MONDAY

Accuracy _____/10 ☆

| 1 | 2 | 3 | 4 | 5 | 6 | 7 | 8 | 9 | 10 |
|---|---|---|---|---|---|---|---|---|---|
|    |    |    |    |    |    | 99 | 24 | 45 | 32 |
| 63 | 52 | 54 | 22 | 45 | 130 | - 33 | 22 | - 21 | 35 |
| - 13 | 32 | 31 | 13 | 23 | 210 | - 40 | - 46 | 03 | - 04 |
| 01 | - 44 | - 20 | - 34 | - 40 | - 30 | 30 | 51 | - 14 | - 60 |
| - 41 | - 40 | - 40 | 02 | - 06 | - 10 | 12 | 25 | 80 | 51 |
|    |    |    |    |    |    |    |    |    |    |

*19:11  © SAI Speed Math Academy, USA*

## DAY 2 – TUESDAY

Accuracy _____/10 ☆

| 1 | 2 | 3 | 4 | 5 | 6 | 7 | 8 | 9 | 10 |
|---|---|---|---|---|---|---|---|---|---|
|    |    |    |    |    |    | 11 | 31 | 15 | 32 |
| 54 | 15 | 55 | 22 | 35 | 250 | 34 | 40 | - 11 | 15 |
| - 13 | 30 | 24 | 22 | 40 | 20 | 22 | - 21 | 53 | - 43 |
| 25 | 54 | - 39 | 33 | - 13 | - 110 | 11 | - 20 | - 30 | 22 |
| - 16 | - 55 | 16 | - 44 | - 50 | - 20 | - 77 | 15 | 70 | - 11 |
|    |    |    |    |    |    |    |    |    |    |

*19:12  © SAI Speed Math Academy, USA*

## DAY 3 – WEDNESDAY

Accuracy _____/10 ☆

| 1 | 2 | 3 | 4 | 5 | 6 | 7 | 8 | 9 | 10 |
|---|---|---|---|---|---|---|---|---|---|
|    |    |    |    |    |    | 89 | 23 | 34 | 14 |
| 15 | 59 | 44 | 11 | 170 | 143 | - 47 | 66 | 61 | 22 |
| - 04 | - 30 | 12 | 32 | - 30 | 20 | - 30 | - 49 | - 12 | - 14 |
| 32 | 20 | - 20 | 14 | 13 | - 13 | 05 | 20 | - 62 | 11 |
| - 40 | - 34 | - 03 | - 30 | - 140 | 105 | - 04 | 15 | 70 | 55 |
|    |    |    |    |    |    |    |    |    |    |

*19:13  © SAI Speed Math Academy, USA*

Accuracy _____/10

| 1 | 2 | 3 | 4 | 5 | 6 | 7 | 8 | 9 | 10 |
|---|---|---|---|---|---|---|---|---|---|
|  |  |  |  |  |  | 55 | 52 | 85 | 46 |
| 25 | 54 | 43 | 96 | 140 | 155 | - 41 | 33 | - 43 | 31 |
| - 12 | 13 | 32 | - 32 | 110 | - 33 | 53 | - 41 | 12 | - 43 |
| 36 | - 41 | - 44 | - 41 | - 20 | 10 | - 42 | - 41 | - 13 | 15 |
| - 40 | - 24 | 35 | - 12 | 150 | - 22 | 50 | 05 | 54 | - 48 |
|  |  |  |  |  |  |  |  |  |  |

19:14  © SAI Speed Math Academy, USA

DAY 5 – FRIDAY

Accuracy _____/10

| 1 | 2 | 3 | 4 | 5 | 6 | 7 | 8 | 9 | 10 |
|---|---|---|---|---|---|---|---|---|---|
|  |  | 21 | 85 | 51 | 54 | 55 | 42 | 66 | 62 |
| 215 | 199 | 34 | - 42 | - 10 | - 21 | - 11 | 43 | - 44 | 10 |
| 31 | - 44 | - 12 | 16 | 05 | 43 | 20 | - 21 | 22 | - 31 |
| - 144 | - 44 | 05 | - 40 | - 44 | - 34 | - 31 | - 34 | - 14 | 12 |
| 33 | 400 | - 24 | - 09 | 93 | 11 | - 23 | 46 | 39 | - 10 |
|  |  |  |  |  |  |  |  |  |  |

19:15  © SAI Speed Math Academy, USA

## SHAPE PUZZLE

Find out what each shape is worth.   Color the shapes.

# WEEK 20 – SKILL BUILDING

## WEEK 20 – PRACTICE WORK

Use Abacus

### DAY 1 – MONDAY

TIME: _____min _____sec     Accuracy _____/16

| 1 | 2 | 3 | 4 | 5 | 6 | 7 | 8 |
|---|---|---|---|---|---|---|---|
| 47 | 31 | 65 | 548 | 549 | 365 | 135 | 145 |
| - 34 | 35 | - 04 | - 144 | 210 | 411 | 461 | 312 |
| 55 | - 41 | 11 | 461 | - 440 | - 323 | - 104 | 412 |
| - 43 | - 25 | - 42 | - 643 | - 207 | - 123 | - 382 | - 258 |
|  |  |  |  |  |  |  |  |

20:1

| 1 | 2 | 3 | 4 | 5 | 6 | 7 | 8 |
|---|---|---|---|---|---|---|---|
|  |  | 25 | 14 | 111 | 653 | 233 | 645 |
| 474 | 148 | 54 | 13 | 23 | 134 | 721 | - 501 |
| 124 | 641 | - 62 | 41 | 544 | - 513 | - 832 | 52 |
| - 42 | - 167 | 42 | - 17 | - 52 | - 71 | 406 | - 132 |
| - 146 | - 310 | - 29 | 28 | 321 | 625 | - 501 | - 64 |
|  |  |  |  |  |  |  |  |

20:2

### DAY 2 – TUESDAY

TIME: _____min _____sec     Accuracy _____/16

| 1 | 2 | 3 | 4 | 5 | 6 | 7 | 8 |
|---|---|---|---|---|---|---|---|
| 44 | 86 | 61 | 577 | 23 | 284 | 341 | 242 |
| 23 | - 55 | 34 | - 303 | 643 | 204 | 346 | 534 |
| - 35 | 12 | - 42 | 601 | - 12 | - 313 | - 450 | - 32 |
| 64 | 21 | - 21 | - 113 | - 141 | - 112 | 321 | - 32 |
|  |  |  |  |  |  |  |  |

20:3

| 1 | 2 | 3 | 4 | 5 | 6 | 7 | 8 |
|---|---|---|---|---|---|---|---|
|  |  | 56 | 54 | 155 | 724 | 555 | 968 |
| 824 | 247 | - 22 | 15 | 123 | 105 | - 421 | - 434 |
| 31 | 32 | 45 | - 41 | - 46 | - 625 | 505 | 200 |
| 130 | - 58 | - 21 | - 17 | 152 | 551 | - 426 | 241 |
| - 245 | 124 | - 52 | 66 | 505 | - 402 | 342 | - 420 |
|  |  |  |  |  |  |  |  |

20:4

## DAY 3 – WEDNESDAY

| 1 | 2 | 3 | 4 | 5 | 6 | 7 | 8 |
|---|---|---|---|---|---|---|---|
| 21 | 42 | 15 | 522 | 462 | 317 | 423 | 831 |
| 44 | 12 | 23 | 134 | 101 | 322 | 443 | 44 |
| - 52 | - 41 | 40 | - 413 | - 202 | 330 | 132 | 113 |
| 34 | 32 | - 34 | 235 | 316 | - 448 | - 663 | - 728 |
|  |  |  |  |  |  |  |  |

20:5

| 1 | 2 | 3 | 4 | 5 | 6 | 7 | 8 |
|---|---|---|---|---|---|---|---|
|  |  | 43 | 369 | 192 | 897 | 446 | 124 |
| 217 | 21 | 11 | 610 | 306 | - 361 | 120 | 252 |
| 361 | 154 | 34 | - 107 | 401 | 121 | 23 | - 31 |
| - 310 | 123 | - 52 | - 130 | - 515 | - 207 | - 146 | - 21 |
| - 135 | - 44 | 33 | - 640 | - 41 | 35 | - 403 | - 103 |
|  |  |  |  |  |  |  |  |

20:6

## DAY 4 – THURSDAY

| 1 | 2 | 3 | 4 | 5 | 6 | 7 | 8 |
|---|---|---|---|---|---|---|---|
| 87 | 29 | 27 | 41 | 66 | 53 | 51 | 84 |
| 12 | 70 | 21 | 33 | 21 | 22 | 34 | 01 |
| - 33 | - 49 | 50 | 15 | - 45 | - 34 | - 25 | - 55 |
| - 55 | 14 | - 44 | - 11 | 54 | - 40 | 17 | 41 |
| 34 | 25 | - 12 | - 15 | - 65 | 56 | - 43 | 15 |
|  |  |  |  |  |  |  |  |

20:7

| 1 | 2 | 3 | 4 | 5 | 6 | 7 | 8 |
|---|---|---|---|---|---|---|---|
| 105 | 246 | 568 | 395 | 561 | 215 | 417 | 45 |
| 852 | 512 | - 241 | - 120 | 407 | 781 | 461 | 110 |
| - 913 | - 204 | 70 | 403 | - 325 | - 983 | 20 | - 41 |
| 351 | - 12 | - 14 | - 46 | - 112 | 154 | - 31 | 215 |
| - 72 | - 102 | - 142 | - 401 | - 130 | - 21 | - 564 | - 123 |
|  |  |  |  |  |  |  |  |

20:8

| 1 | 2 | 3 | 4 | 5 | 6 | 7 | 8 |
|---|---|---|---|---|---|---|---|
| 41 | 61 | 65 | 78 | 72 | 38 | 87 | 21 |
| 45 | 13 | 31 | 20 | 14 | 21 | - 24 | 58 |
| - 13 | - 43 | - 55 | - 44 | - 31 | - 56 | 11 | - 23 |
| 21 | - 21 | 33 | - 20 | - 12 | 24 | 21 | - 53 |
| - 44 | 59 | 11 | 10 | 40 | - 17 | - 54 | 14 |
|  |  |  |  |  |  |  |  |

20:9

| 1 | 2 | 3 | 4 | 5 | 6 | 7 | 8 |
|---|---|---|---|---|---|---|---|
| 654 | 214 | 56 | 162 | 202 | 251 | 133 | 345 |
| - 250 | 55 | 201 | 135 | 103 | 14 | 221 | - 34 |
| - 101 | - 136 | 402 | - 13 | 453 | 222 | 344 | 21 |
| 46 | 55 | - 109 | - 231 | - 12 | - 31 | - 423 | - 32 |
| - 208 | - 112 | - 310 | - 22 | - 513 | - 445 | 511 | 244 |
|  |  |  |  |  |  |  |  |

20:10

## WEEK 20 – DICTATION

**DO 6 PROBLEMS A DAY** and write your answers below.

Students: Try calculating all dictation problems in mind.

Teachers: Dictate any two digit problem from this week's homework.

| 1 | 2 | 3 | 4 | 5 | 6 | 7 | 8 | 9 | 10 |
|---|---|---|---|---|---|---|---|---|---|
|  |  |  |  |  |  |  |  |  |  |

| 11 | 12 | 13 | 14 | 15 | 16 | 17 | 18 | 19 | 20 |
|---|---|---|---|---|---|---|---|---|---|
|  |  |  |  |  |  |  |  |  |  |

| 21 | 22 | 23 | 24 | 25 | 26 | 27 | 28 | 29 | 30 |
|---|---|---|---|---|---|---|---|---|---|
|  |  |  |  |  |  |  |  |  |  |

# WEEK 20 – MIND MATH PRACTICE

Visualize

## DAY 1 – MONDAY   Accuracy _____/10 ☆

| 1 | 2 | 3 | 4 | 5 | 6 | 7 | 8 | 9 | 10 |
|---|---|---|---|---|---|---|---|---|---|
|    |    |    |     |     | 24 | 05 | 12 | 44 | 54 |
| 41 | 32 | 13 | 202 | 131 | 23 | 04 | 41 | 11 | 31 |
| 13 | 24 | 24 | 63 | 44 | - 04 | 50 | 11 | - 25 | - 62 |
| 12 | 20 | 30 | - 152 | - 51 | 30 | 40 | 12 | 33 | 26 |
| - 41 | - 55 | - 17 | 45 | - 100 | 05 | - 61 | - 21 | 22 | - 12 |
|    |    |    |     |     |    |    |    |    |    |

20:11

## DAY 2 – TUESDAY   Accuracy _____/10 ☆

| 1 | 2 | 3 | 4 | 5 | 6 | 7 | 8 | 9 | 10 |
|---|---|---|---|---|---|---|---|---|---|
|    |    |    |     |     | 22 | 55 | 51 | 99 | 88 |
| 12 | 23 | 68 | 272 | 510 | 74 | - 25 | 13 | - 43 | - 45 |
| 32 | 12 | - 21 | - 32 | - 100 | - 35 | 52 | - 24 | - 32 | 52 |
| - 24 | 21 | - 21 | 520 | 230 | 16 | - 41 | 12 | 25 | - 63 |
| 37 | - 04 | - 02 | - 50 | - 440 | - 41 | 52 | 17 | 10 | 35 |
|    |    |    |     |     |    |    |    |    |    |

20:12

## DAY 3 – WEDNESDAY   Accuracy _____/10 ☆

| 1 | 2 | 3 | 4 | 5 | 6 | 7 | 8 | 9 | 10 |
|---|---|---|---|---|---|---|---|---|---|
|    |    |    |    |     |     | 22 | 59 | 66 | 33 |
| 11 | 66 | 25 | 54 | 166 | 151 | 25 | - 26 | 11 | 44 |
| 24 | - 24 | 22 | 34 | - 23 | 130 | - 03 | 44 | - 22 | - 12 |
| 11 | - 11 | - 31 | - 13 | 10 | - 240 | 50 | - 12 | - 13 | 30 |
| 41 | 46 | 42 | - 21 | - 110 | 51 | - 20 | 21 | - 10 | - 55 |
|    |    |    |    |     |     |    |    |    |    |

20:13

www.abacus-math.com

| 1 | 2 | 3 | 4 | 5 | 6 | 7 | 8 | 9 | 10 |
|---|---|---|---|---|---|---|---|---|---|
|  |  |  |  | 31 | 32 | 39 | 85 | 85 | 72 |
| 37 | 211 | 183 | 462 | 45 | 56 | 20 | - 43 | - 64 | 17 |
| 201 | 40 | - 51 | 15 | - 55 | - 64 | - 53 | 15 | 45 | - 48 |
| - 104 | - 20 | 44 | - 56 | 76 | 23 | 61 | 31 | - 32 | 16 |
| 01 | - 120 | - 150 | - 21 | - 64 | 22 | - 24 | - 76 | 11 | - 55 |
|  |  |  |  |  |  |  |  |  |  |

20:14

| 1 | 2 | 3 | 4 | 5 | 6 | 7 | 8 | 9 | 10 |
|---|---|---|---|---|---|---|---|---|---|
|  |  |  |  |  | 88 | 45 | 23 | 44 | 25 |
| 120 | 62 | 25 | 95 | 115 | - 15 | - 13 | 13 | 41 | 44 |
| 120 | 106 | 54 | - 41 | 130 | - 51 | 31 | 43 | - 31 | - 23 |
| 12 | - 40 | 110 | 122 | - 202 | - 11 | - 22 | - 12 | - 33 | - 16 |
| - 200 | - 18 | - 149 | - 66 | 55 | 52 | 50 | - 13 | 55 | 19 |
|  |  |  |  |  |  |  |  |  |  |

20:15

## SHAPE PUZZLE

Find out what each shape is worth. Color the shapes.

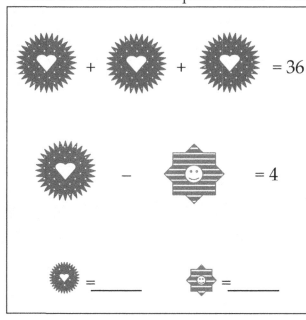

=____     =____     =____     =____

# END OF LEVEL 1 – TEST

## CALCULATE ON ABACUS

AIM FOR ACCURACY

Use Abacus

| 1 | 2 | 3 | 4 | 5 | 6 | 7 | 8 | 9 | 10 | |
|---|---|---|---|---|---|---|---|---|---|---|
| 11 | 22 | 13 | 12 | 10 | 11 | 24 | 23 | 41 | 23 | Test:1 |
| 13 | 12 | 23 | 23 | 41 | 43 | 11 | 34 | 13 | 44 | |
| 21 | 21 | 32 | 34 | 34 | 34 | - 21 | - 21 | - 52 | - 21 | |
| 12 | 22 | 30 | 10 | 12 | 10 | 13 | 33 | 44 | 42 | |
| | | | | | | | | | | |

| 1 | 2 | 3 | 4 | 5 | 6 | 7 | 8 | 9 | 10 | |
|---|---|---|---|---|---|---|---|---|---|---|
| 55 | 13 | 44 | 21 | 34 | 56 | 24 | 25 | 35 | 33 | Test:2 |
| - 12 | 43 | 41 | 34 | 31 | - 22 | 23 | 23 | 34 | 42 | |
| 31 | - 13 | - 31 | - 23 | - 32 | 31 | 41 | 40 | - 23 | - 31 | |
| - 22 | - 13 | 21 | - 31 | 21 | - 40 | - 64 | - 42 | - 14 | 21 | |
| | | | | | | | | | | |

| 1 | 2 | 3 | 4 | 5 | 6 | 7 | 8 | 9 | 10 | |
|---|---|---|---|---|---|---|---|---|---|---|
| 43 | 63 | 62 | 12 | 97 | 22 | 56 | 34 | 14 | 24 | Test:3 |
| 43 | - 21 | 24 | 42 | - 32 | 34 | - 13 | 41 | 24 | 31 | |
| - 24 | 34 | - 31 | - 32 | - 33 | - 43 | - 13 | - 23 | 31 | - 22 | |
| 13 | - 31 | - 13 | 53 | - 11 | 41 | 41 | - 11 | - 48 | - 13 | |
| | | | | | | | | | | |

| 1 | 2 | 3 | 4 | 5 | 6 | 7 | 8 | 9 | 10 | |
|---|---|---|---|---|---|---|---|---|---|---|
| 68 | 75 | 14 | 44 | 66 | 77 | 85 | 36 | 43 | 51 | Test:4 |
| - 34 | - 34 | 31 | 31 | - 14 | 21 | - 34 | 12 | 21 | 34 | |
| 41 | 11 | - 24 | - 43 | 33 | - 42 | - 41 | - 44 | - 41 | - 42 | |
| - 33 | 14 | 32 | - 12 | - 11 | - 43 | 25 | 12 | 15 | - 13 | |
| | | | | | | | | | | |

www.abacus-math.com

| *1* | *2* | *3* | *4* | *5* | *6* | *7* | *8* | *9* | *10* | |
|---|---|---|---|---|---|---|---|---|---|---|
| 25 | 54 | 43 | 34 | 46 | 55 | 55 | 52 | 58 | 64 | Test:5 |
| - 11 | 13 | 31 | 32 | 11 | - 42 | - 42 | - 31 | - 43 | 23 | |
| 32 | - 41 | - 44 | - 41 | - 24 | 12 | 33 | 41 | 32 | - 42 | |
| - 24 | - 24 | 31 | - 12 | - 32 | - 13 | - 24 | - 21 | - 13 | - 12 | |
| | | | | | | | | | | |

| *1* | *2* | *3* | *4* | *5* | *6* | *7* | *8* | *9* | *10* | |
|---|---|---|---|---|---|---|---|---|---|---|
| 97 | 65 | 54 | 43 | 35 | 43 | 75 | 44 | 64 | 45 | Test:6 |
| - 34 | - 45 | 22 | 22 | - 23 | 33 | - 31 | 31 | 32 | 24 | |
| - 31 | 32 | - 34 | - 42 | 64 | - 42 | - 24 | - 32 | - 41 | - 45 | |
| 34 | 34 | 12 | 43 | - 43 | 11 | 30 | 24 | - 42 | 31 | |
| | | | | | | | | | | |

| *1* | *2* | *3* | *4* | *5* | *6* | *7* | *8* | *9* | *10* | |
|---|---|---|---|---|---|---|---|---|---|---|
| 25 | 12 | 11 | 73 | 11 | 33 | 21 | 41 | 42 | 52 | Test:7 |
| 12 | 43 | 40 | - 12 | 34 | 44 | 43 | 34 | 42 | 43 | |
| 31 | 31 | 11 | 14 | 31 | - 21 | - 31 | - 21 | - 61 | - 21 | |
| - 44 | - 43 | - 42 | - 64 | 13 | - 31 | 21 | 12 | 13 | - 41 | |
| 21 | 20 | 67 | 32 | - 48 | - 21 | 23 | - 44 | 40 | 33 | |
| | | | | | | | | | | |

| *1* | *2* | *3* | *4* | *5* | *6* | *7* | *8* | *9* | *10* | |
|---|---|---|---|---|---|---|---|---|---|---|
| 34 | 35 | 66 | 12 | 53 | 53 | 42 | 25 | 74 | 22 | Test:8 |
| - 13 | 31 | 21 | 23 | 14 | 21 | 34 | 23 | - 31 | 13 | |
| 25 | - 54 | - 32 | 21 | - 13 | - 13 | - 25 | - 44 | - 23 | - 34 | |
| - 34 | 41 | 13 | - 44 | - 12 | - 21 | 16 | - 03 | 14 | 86 | |
| 75 | - 40 | - 48 | - 02 | 30 | 30 | 32 | 50 | 51 | - 54 | |
| | | | | | | | | | | |

# DICTATION

**Calculate dictation problems on abacus.**

| *1* | *2* | *3* | *4* | *5* | *6* | *7* | *8* | *9* | *10* |
|---|---|---|---|---|---|---|---|---|---|
| | | | | | | | | | |

# MIND MATH

| 1 | 2 | 3 | 4 | 5 | 6 | 7 | 8 | 9 | 10 | |
|---|---|---|---|---|---|---|---|---|---|---|
| 11 | 22 | 13 | 14 | 05 | 12 | 44 | 54 | 41 | 31 | |
| 13 | 24 | 24 | 13 | 04 | 41 | 11 | 31 | 13 | 44 | |
| 11 | 20 | 30 | - 04 | 50 | 11 | - 25 | - 61 | - 52 | - 51 | Test:9 |
| 41 | - 55 | - 17 | 30 | 40 | 12 | 33 | 23 | 44 | 13 | |
|   |   |   |   |   |   |   |   |   |   | |

| 1 | 2 | 3 | 4 | 5 | 6 | 7 | 8 | 9 | 10 | |
|---|---|---|---|---|---|---|---|---|---|---|
| 12 | 23 | 68 | 74 | 55 | 51 | 99 | 88 | 66 | 51 | |
| 32 | 22 | - 31 | - 34 | - 25 | 13 | - 44 | - 33 | - 33 | - 10 | |
| - 24 | 21 | - 11 | 16 | 52 | - 24 | - 33 | - 22 | 22 | 23 | Test:10 |
| 37 | - 04 | - 02 | - 41 | - 41 | 12 | 22 | - 22 | - 14 | - 44 | |
|   |   |   |   |   |   |   |   |   |   | |

# DICTATION

**Calculate dictation problems in mind.**

| 1 | 2 | 3 | 4 | 5 | 6 | 7 | 8 | 9 | 10 |
|---|---|---|---|---|---|---|---|---|---|
|   |   |   |   |   |   |   |   |   |   |

♥ + ⚑ = 12

♥ – ⚑ = 04

♥ = _____   ⚑ = _____

◯ – ♥ = 2

◯ + ◯ – ☾ = 11

◯ = _____   ☾ = _____

# ANSWER KEY

## WEEK 11 – PRACTICE WORK
### DAY 1 – MONDAY

| 1 | 2 | 3 | 4 | 5 | 6 | 7 | 8 | |
|---|---|---|---|---|---|---|---|---|
| 84 | 56 | 51 | 69 | 44 | 68 | 01 | 95 | 11:1 |

| 1 | 2 | 3 | 4 | 5 | 6 | 7 | 8 | |
|---|---|---|---|---|---|---|---|---|
| 66 | 75 | 11 | 79 | 67 | 44 | 06 | 69 | 11:2 |

| 1 | 2 | 3 | 4 | 5 | 6 | 7 | 8 | |
|---|---|---|---|---|---|---|---|---|
| 54 | 99 | 16 | 15 | 40 | 21 | 40 | 41 | 11:3 |

| 1 | 2 | 3 | 4 | 5 | 6 | 7 | 8 | |
|---|---|---|---|---|---|---|---|---|
| 51 | 10 | 02 | 56 | 96 | 05 | 15 | 96 | 11:4 |

### DAY 2 – TUESDAY

| 1 | 2 | 3 | 4 | 5 | 6 | 7 | 8 | |
|---|---|---|---|---|---|---|---|---|
| 98 | 76 | 95 | 15 | 85 | 12 | 94 | 66 | 11:5 |

| 1 | 2 | 3 | 4 | 5 | 6 | 7 | 8 | |
|---|---|---|---|---|---|---|---|---|
| 52 | 99 | 44 | 84 | 17 | 67 | 14 | 35 | 11:6 |

### DAY 3 – WEDNESDAY

| 1 | 2 | 3 | 4 | 5 | 6 | 7 | 8 | |
|---|---|---|---|---|---|---|---|---|
| 30 | 55 | 00 | 44 | 51 | 66 | 41 | 00 | 11:7 |

| 1 | 2 | 3 | 4 | 5 | 6 | 7 | 8 | |
|---|---|---|---|---|---|---|---|---|
| 68 | 88 | 47 | 21 | 08 | 14 | 00 | 04 | 11:8 |

### DAY 4 – THURSDAY

| 1 | 2 | 3 | 4 | 5 | 6 | 7 | 8 | |
|---|---|---|---|---|---|---|---|---|
| 66 | 75 | 04 | 55 | 54 | 12 | 00 | 16 | 11:9 |

| 1 | 2 | 3 | 4 | 5 | 6 | 7 | 8 | |
|---|---|---|---|---|---|---|---|---|
| 98 | 01 | 20 | 86 | 41 | 40 | 96 | 40 | 11:10 |

### DAY 5 – FRIDAY

| 1 | 2 | 3 | 4 | 5 | 6 | 7 | 8 | |
|---|---|---|---|---|---|---|---|---|
| 81 | 69 | 00 | 66 | 11 | 67 | 86 | 02 | 11:11 |

| 1 | 2 | 3 | 4 | 5 | 6 | 7 | 8 | |
|---|---|---|---|---|---|---|---|---|
| 34 | 66 | 65 | 55 | 00 | 50 | 59 | 98 | 11:12 |

## WEEK 11 – MIND MATH PRACTICE WORK
### DAY 1 – MONDAY

| 1 | 2 | 3 | 4 | 5 | 6 | 7 | 8 | 9 | 10 | |
|---|---|---|---|---|---|---|---|---|----|---|
| 00 | 57 | 50 | 64 | 25 | 65 | 05 | 40 | 46 | 78 | 11:13 |

### DAY 2 – TUESDAY

| 1 | 2 | 3 | 4 | 5 | 6 | 7 | 8 | 9 | 10 | |
|---|---|---|---|---|---|---|---|---|----|---|
| 35 | 46 | 26 | 54 | 55 | 11 | 24 | 51 | 34 | 41 | 11:14 |

### DAY 3 – WEDNESDAY

| 1 | 2 | 3 | 4 | 5 | 6 | 7 | 8 | 9 | 10 | |
|---|---|---|---|---|---|---|---|---|----|---|
| 36 | 46 | 65 | 26 | 64 | 12 | 72 | 88 | 44 | 03 | 11:15 |

### DAY 4 – THURSDAY

| 1 | 2 | 3 | 4 | 5 | 6 | 7 | 8 | 9 | 10 | |
|---|---|---|---|---|---|---|---|---|----|---|
| 64 | 44 | 22 | 04 | 40 | 11 | 94 | 45 | 33 | 44 | 11:16 |

### DAY 5 – FRIDAY

| 1 | 2 | 3 | 4 | 5 | 6 | 7 | 8 | 9 | 10 | |
|---|---|---|---|---|---|---|---|---|----|---|
| 66 | 51 | 65 | 56 | 09 | 05 | 51 | 69 | 15 | 52 | 11:17 |

| 1 | 2 | 3 | 4 | 5 | 6 | 7 | 8 | 9 | 10 | |
|---|---|---|---|---|---|---|---|---|----|---|
| 46 | 06 | 61 | 51 | 35 | 84 | 16 | 46 | 01 | 23 | 11:18 |

### WEEK 11 – SHAPE PUZZLES

▭ = 12     △ = 6

◡ = 10     ▤ = 9

## WEEK 12 – PRACTICE WORK

### DAY 1 – MONDAY

| 1 | 2 | 3 | 4 | 5 | 6 | 7 | 8 | |
|---|---|---|---|---|---|---|---|---|
| 18 | 46 | 33 | 00 | 04 | 40 | 46 | 37 | 12:1 |

| 1 | 2 | 3 | 4 | 5 | 6 | 7 | 8 | |
|---|---|---|---|---|---|---|---|---|
| 43 | 24 | 40 | 33 | 14 | 68 | 11 | 48 | 12:2 |

| 1 | 2 | 3 | 4 | 5 | 6 | 7 | 8 | |
|---|---|---|---|---|---|---|---|---|
| 20 | 58 | 83 | 01 | 49 | 78 | 11 | 12 | 12:3 |

| 1 | 2 | 3 | 4 | 5 | 6 | 7 | 8 | |
|---|---|---|---|---|---|---|---|---|
| 58 | 20 | 96 | 98 | 10 | 42 | 31 | 33 | 12:4 |

### DAY 2 – TUESDAY

| 1 | 2 | 3 | 4 | 5 | 6 | 7 | 8 | |
|---|---|---|---|---|---|---|---|---|
| 00 | 44 | 12 | 76 | 01 | 31 | 41 | 23 | 12:5 |

| 1 | 2 | 3 | 4 | 5 | 6 | 7 | 8 | |
|---|---|---|---|---|---|---|---|---|
| 40 | 54 | 04 | 44 | 32 | 10 | 58 | 32 | 12:6 |

### DAY 3 – WEDNESDAY

| 1 | 2 | 3 | 4 | 5 | 6 | 7 | 8 | |
|---|---|---|---|---|---|---|---|---|
| 98 | 13 | 21 | 59 | 01 | 03 | 30 | 21 | 12:7 |

| 1 | 2 | 3 | 4 | 5 | 6 | 7 | 8 | |
|---|---|---|---|---|---|---|---|---|
| 58 | 22 | 54 | 99 | 41 | 03 | 33 | 69 | 12:8 |

### DAY 4 – THURSDAY

| 1 | 2 | 3 | 4 | 5 | 6 | 7 | 8 | |
|---|---|---|---|---|---|---|---|---|
| 75 | 59 | 84 | 38 | 55 | 44 | 95 | 83 | 12:9 |

| 1 | 2 | 3 | 4 | 5 | 6 | 7 | 8 | |
|---|---|---|---|---|---|---|---|---|
| 59 | 49 | 42 | 28 | 46 | 89 | 04 | 03 | 12:10 |

### DAY 5 – FRIDAY

| 1 | 2 | 3 | 4 | 5 | 6 | 7 | 8 | |
|---|---|---|---|---|---|---|---|---|
| 76 | 34 | 55 | 76 | 65 | 54 | 44 | 37 | 12:11 |

| 1 | 2 | 3 | 4 | 5 | 6 | 7 | 8 | |
|---|---|---|---|---|---|---|---|---|
| 88 | 30 | 75 | 58 | 44 | 69 | 13 | 11 | 12:12 |

## WEEK 12 – MIND MATH PRACTICE WORK

### DAY 1 – MONDAY

| 1 | 2 | 3 | 4 | 5 | 6 | 7 | 8 | 9 | 10 | |
|---|---|---|---|---|---|---|---|---|---|---|
| 03 | 57 | 01 | 42 | 03 | 24 | 50 | 01 | 20 | 33 | 12:13 |

### DAY 2 – TUESDAY

| 1 | 2 | 3 | 4 | 5 | 6 | 7 | 8 | 9 | 10 | |
|---|---|---|---|---|---|---|---|---|---|---|
| 15 | 55 | 03 | 53 | 23 | 13 | 33 | 34 | 25 | 34 | 12:14 |

### DAY 3 – WEDNESDAY

| 1 | 2 | 3 | 4 | 5 | 6 | 7 | 8 | 9 | 10 | |
|---|---|---|---|---|---|---|---|---|---|---|
| 60 | 25 | 26 | 30 | 34 | 51 | 56 | 00 | 40 | 89 | 12:15 |

### DAY 4 – THURSDAY

| 1 | 2 | 3 | 4 | 5 | 6 | 7 | 8 | 9 | 10 | |
|---|---|---|---|---|---|---|---|---|---|---|
| 40 | 36 | 28 | 94 | 43 | 03 | 30 | 31 | 40 | 43 | 12:16 |

### DAY 5 – FRIDAY

| 1 | 2 | 3 | 4 | 5 | 6 | 7 | 8 | 9 | 10 | |
|---|---|---|---|---|---|---|---|---|---|---|
| 33 | 22 | 33 | 54 | 41 | 50 | 93 | 33 | 56 | 22 | 12:17 |

| 1 | 2 | 3 | 4 | 5 | 6 | 7 | 8 | 9 | 10 | |
|---|---|---|---|---|---|---|---|---|---|---|
| 00 | 00 | 89 | 20 | 33 | 43 | 40 | 99 | 03 | 00 | 12:18 |

WEEK 12 – SHAPE PUZZLES

☆ = 11    ▨ = 9

☼ = 7    ♜ = 4

## WEEK 13 – PRACTICE WORK

### DAY 1 – MONDAY

| 1 | 2 | 3 | 4 | 5 | 6 | 7 | 8 | |
|---|---|---|---|---|---|---|---|---|
| 59 | 72 | 77 | 63 | 88 | 59 | 61 | 85 | 13:1 |

| 1 | 2 | 3 | 4 | 5 | 6 | 7 | 8 | |
|---|---|---|---|---|---|---|---|---|
| 78 | 55 | 15 | 55 | 76 | 14 | 62 | 96 | 13:2 |

| 1 | 2 | 3 | 4 | 5 | 6 | 7 | 8 | |
|---|---|---|---|---|---|---|---|---|
| 49 | 77 | 62 | 75 | 97 | 78 | 27 | 44 | 13:3 |

| 1 | 2 | 3 | 4 | 5 | 6 | 7 | 8 | |
|---|---|---|---|---|---|---|---|---|
| 42 | 11 | 69 | 64 | 04 | 54 | 97 | 41 | 13:4 |

### DAY 2 – TUESDAY

| 1 | 2 | 3 | 4 | 5 | 6 | 7 | 8 | |
|---|---|---|---|---|---|---|---|---|
| 66 | 99 | 43 | 15 | 06 | 78 | 97 | 18 | 13:5 |

| 1 | 2 | 3 | 4 | 5 | 6 | 7 | 8 | |
|---|---|---|---|---|---|---|---|---|
| 46 | 54 | 99 | 59 | 31 | 27 | 45 | 04 | 13:6 |

### DAY 3 – WEDNESDAY

| 1 | 2 | 3 | 4 | 5 | 6 | 7 | 8 | |
|---|---|---|---|---|---|---|---|---|
| 32 | 34 | 51 | 14 | 43 | 45 | 97 | 34 | 13:7 |

| 1 | 2 | 3 | 4 | 5 | 6 | 7 | 8 | |
|---|---|---|---|---|---|---|---|---|
| 39 | 01 | 66 | 79 | 84 | 21 | 67 | 58 | 13:8 |

### DAY 4 – THURSDAY

| 1 | 2 | 3 | 4 | 5 | 6 | 7 | 8 | |
|---|---|---|---|---|---|---|---|---|
| 15 | 42 | 28 | 87 | 66 | 75 | 45 | 10 | 13:9 |

| 1 | 2 | 3 | 4 | 5 | 6 | 7 | 8 | |
|---|---|---|---|---|---|---|---|---|
| 76 | 40 | 79 | 54 | 76 | 40 | 00 | 03 | 13:10 |

### DAY 5 – FRIDAY

| 1 | 2 | 3 | 4 | 5 | 6 | 7 | 8 | |
|---|---|---|---|---|---|---|---|---|
| 24 | 04 | 40 | 14 | 15 | 56 | 55 | 11 | 13:11 |

| 1 | 2 | 3 | 4 | 5 | 6 | 7 | 8 | |
|---|---|---|---|---|---|---|---|---|
| 14 | 54 | 00 | 33 | 41 | 69 | 70 | 01 | 13:12 |

## WEEK 13 – MIND MATH PRACTICE WORK

### DAY 1 – MONDAY

| 1 | 2 | 3 | 4 | 5 | 6 | 7 | 8 | 9 | 10 | |
|---|---|---|---|---|---|---|---|---|----|---|
| 11 | 01 | 06 | 42 | 00 | 47 | 76 | 53 | 27 | 54 | 13:13 |

### DAY 2 – TUESDAY

| 1 | 2 | 3 | 4 | 5 | 6 | 7 | 8 | 9 | 10 | |
|---|---|---|---|---|---|---|---|---|----|---|
| 64 | 26 | 37 | 50 | 65 | 47 | 44 | 21 | 34 | 24 | 13:14 |

### DAY 3 – WEDNESDAY

| 1 | 2 | 3 | 4 | 5 | 6 | 7 | 8 | 9 | 10 | |
|---|---|---|---|---|---|---|---|---|----|---|
| 09 | 05 | 55 | 71 | 17 | 05 | 16 | 85 | 65 | 55 | 13:15 |

### DAY 4 – THURSDAY

| 1 | 2 | 3 | 4 | 5 | 6 | 7 | 8 | 9 | 10 | |
|---|---|---|---|---|---|---|---|---|----|---|
| 53 | 20 | 75 | 36 | 66 | 37 | 99 | 64 | 52 | 65 | 13:16 |

### DAY 5 – FRIDAY

| 1 | 2 | 3 | 4 | 5 | 6 | 7 | 8 | 9 | 10 | |
|---|---|---|---|---|---|---|---|---|----|---|
| 65 | 45 | 55 | 35 | 15 | 66 | 54 | 53 | 39 | 77 | 13:17 |

| 1 | 2 | 3 | 4 | 5 | 6 | 7 | 8 | 9 | 10 | |
|---|---|---|---|---|---|---|---|---|----|---|
| 06 | 54 | 55 | 45 | 04 | 77 | 37 | 65 | 40 | 32 | 13:18 |

WEEK 13 – SHAPE PUZZLES

= 7      = 14

= 1      = 5

# WEEK 14 – PRACTICE WORK
## DAY 1 – MONDAY

| 1 | 2 | 3 | 4 | 5 | 6 | 7 | 8 | |
|----|----|----|----|----|----|----|----|----|
| 52 | 53 | 28 | 00 | 05 | 23 | 11 | 66 | 14:1 |

| 1 | 2 | 3 | 4 | 5 | 6 | 7 | 8 | |
|----|----|----|----|----|----|----|----|----|
| 14 | 54 | 12 | 59 | 28 | 87 | 62 | 02 | 14:2 |

| 1 | 2 | 3 | 4 | 5 | 6 | 7 | 8 | |
|----|----|----|----|----|----|----|----|----|
| 56 | 46 | 22 | 79 | 10 | 20 | 63 | 24 | 14:3 |

| 1 | 2 | 3 | 4 | 5 | 6 | 7 | 8 | |
|----|----|----|----|----|----|----|----|----|
| 23 | 59 | 11 | 75 | 94 | 74 | 00 | 23 | 14:4 |

## DAY 2 – TUESDAY

| 1 | 2 | 3 | 4 | 5 | 6 | 7 | 8 | |
|----|----|----|----|----|----|----|----|----|
| 15 | 03 | 32 | 08 | 41 | 64 | 54 | 58 | 14:5 |

| 1 | 2 | 3 | 4 | 5 | 6 | 7 | 8 | |
|----|----|----|----|----|----|----|----|----|
| 32 | 53 | 87 | 02 | 12 | 56 | 40 | 89 | 14:6 |

## DAY 3 – WEDNESDAY

| 1 | 2 | 3 | 4 | 5 | 6 | 7 | 8 | |
|----|----|----|----|----|----|----|----|----|
| 21 | 24 | 45 | 32 | 04 | 20 | 95 | 33 | 14:7 |

| 1 | 2 | 3 | 4 | 5 | 6 | 7 | 8 | |
|----|----|----|----|----|----|----|----|----|
| 86 | 01 | 22 | 24 | 42 | 34 | 00 | 20 | 14:8 |

## DAY 4 – THURSDAY

| 1 | 2 | 3 | 4 | 5 | 6 | 7 | 8 | |
|----|----|----|----|----|----|----|----|----|
| 11 | 67 | 21 | 73 | 01 | 20 | 58 | 32 | 14:9 |

| 1 | 2 | 3 | 4 | 5 | 6 | 7 | 8 | |
|----|----|----|----|----|----|----|----|----|
| 02 | 69 | 02 | 23 | 13 | 53 | 03 | 34 | 14:10 |

## DAY 5 – FRIDAY

| 1 | 2 | 3 | 4 | 5 | 6 | 7 | 8 | |
|----|----|----|----|----|----|----|----|----|
| 57 | 05 | 02 | 14 | 00 | 44 | 14 | 95 | 14:11 |

| 1 | 2 | 3 | 4 | 5 | 6 | 7 | 8 | |
|----|----|----|----|----|----|----|----|----|
| 78 | 94 | 92 | 04 | 75 | 13 | 00 | 25 | 14:12 |

# WEEK 14 – MIND MATH PRACTICE WORK
## DAY 1 – MONDAY

| 1 | 2 | 3 | 4 | 5 | 6 | 7 | 8 | 9 | 10 | |
|----|----|----|----|----|----|----|----|----|----|----|
| 30 | 19 | 03 | 90 | 30 | 07 | 30 | 34 | 40 | 10 | 14:13 |

## DAY 2 – TUESDAY

| 1 | 2 | 3 | 4 | 5 | 6 | 7 | 8 | 9 | 10 | |
|----|----|----|----|----|----|----|----|----|----|----|
| 31 | 63 | 20 | 34 | 29 | 26 | 03 | 24 | 01 | 13 | 14:14 |

## DAY 3 – WEDNESDAY

| 1 | 2 | 3 | 4 | 5 | 6 | 7 | 8 | 9 | 10 | |
|----|----|----|----|----|----|----|----|----|----|----|
| 05 | 02 | 33 | 83 | 22 | 21 | 16 | 24 | 34 | 00 | 14:15 |

## DAY 4 – THURSDAY

| 1 | 2 | 3 | 4 | 5 | 6 | 7 | 8 | 9 | 10 | |
|----|----|----|----|----|----|----|----|----|----|----|
| 20 | 32 | 24 | 52 | 44 | 43 | 81 | 00 | 08 | 25 | 14:16 |

## DAY 5 – FRIDAY

| 1 | 2 | 3 | 4 | 5 | 6 | 7 | 8 | 9 | 10 | |
|----|----|----|----|----|----|----|----|----|----|----|
| 22 | 02 | 44 | 33 | 00 | 22 | 30 | 25 | 22 | 21 | 14:17 |

| 1 | 2 | 3 | 4 | 5 | 6 | 7 | 8 | 9 | 10 | |
|----|----|----|----|----|----|----|----|----|----|----|
| 04 | 88 | 46 | 77 | 50 | 98 | 54 | 51 | 40 | 31 | 14:18 |

### WEEK 14 – SHAPE PUZZLES

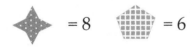

= 8     = 6

= 2     = 4

## WEEK 15 – PRACTICE WORK
### DAY 1 – MONDAY

| 1 | 2 | 3 | 4 | 5 | 6 | 7 | 8 | |
|---|---|---|---|---|---|---|---|---|
| 54 | 10 | 21 | 22 | 52 | 76 | 00 | 71 | 15:1 |

| 1 | 2 | 3 | 4 | 5 | 6 | 7 | 8 | |
|---|---|---|---|---|---|---|---|---|
| 44 | 11 | 30 | 33 | 00 | 54 | 46 | 24 | 15:2 |

### DAY 2 – TUESDAY

| 1 | 2 | 3 | 4 | 5 | 6 | 7 | 8 | |
|---|---|---|---|---|---|---|---|---|
| 11 | 44 | 44 | 00 | 32 | 28 | 77 | 64 | 15:3 |

| 1 | 2 | 3 | 4 | 5 | 6 | 7 | 8 | |
|---|---|---|---|---|---|---|---|---|
| 94 | 52 | 14 | 05 | 53 | 87 | 20 | 01 | 15:4 |

### DAY 3 – WEDNESDAY

| 1 | 2 | 3 | 4 | 5 | 6 | 7 | 8 | |
|---|---|---|---|---|---|---|---|---|
| 30 | 40 | 69 | 42 | 02 | 58 | 00 | 22 | 15:5 |

| 1 | 2 | 3 | 4 | 5 | 6 | 7 | 8 | |
|---|---|---|---|---|---|---|---|---|
| 62 | 24 | 11 | 76 | 93 | 96 | 84 | 45 | 15:6 |

### DAY 4 – THURSDAY

| 1 | 2 | 3 | 4 | 5 | 6 | 7 | 8 | |
|---|---|---|---|---|---|---|---|---|
| 88 | 34 | 86 | 65 | 32 | 01 | 24 | 65 | 15:7 |

| 1 | 2 | 3 | 4 | 5 | 6 | 7 | 8 | |
|---|---|---|---|---|---|---|---|---|
| 44 | 41 | 11 | 68 | 95 | 13 | 53 | 43 | 15:8 |

### DAY 5 – FRIDAY

| 1 | 2 | 3 | 4 | 5 | 6 | 7 | 8 | |
|---|---|---|---|---|---|---|---|---|
| 34 | 31 | 22 | 25 | 24 | 08 | 25 | 22 | 15:9 |

| 1 | 2 | 3 | 4 | 5 | 6 | 7 | 8 | |
|---|---|---|---|---|---|---|---|---|
| 45 | 33 | 20 | 28 | 04 | 48 | 23 | 41 | 15:10 |

## WEEK 15 – MIND MATH PRACTICE WORK
### DAY 1 – MONDAY

| 1 | 2 | 3 | 4 | 5 | 6 | 7 | 8 | 9 | 10 | |
|---|---|---|---|---|---|---|---|---|----|---|
| 46 | 15 | 60 | 14 | 32 | 98 | 12 | 64 | 13 | 70 | 15:11 |

### DAY 2 – TUESDAY

| 1 | 2 | 3 | 4 | 5 | 6 | 7 | 8 | 9 | 10 | |
|---|---|---|---|---|---|---|---|---|----|---|
| 25 | 21 | 00 | 56 | 39 | 53 | 60 | 32 | 03 | 86 | 15:12 |

### DAY 3 – WEDNESDAY

| 1 | 2 | 3 | 4 | 5 | 6 | 7 | 8 | 9 | 10 | |
|---|---|---|---|---|---|---|---|---|----|---|
| 49 | 35 | 63 | 56 | 77 | 94 | 00 | 02 | 31 | 66 | 15:13 |

### DAY 4 – THURSDAY

| 1 | 2 | 3 | 4 | 5 | 6 | 7 | 8 | 9 | 10 | |
|---|---|---|---|---|---|---|---|---|----|---|
| 73 | 21 | 24 | 20 | 44 | 78 | 53 | 77 | 21 | 38 | 15:14 |

### DAY 5 – FRIDAY

| 1 | 2 | 3 | 4 | 5 | 6 | 7 | 8 | 9 | 10 | |
|---|---|---|---|---|---|---|---|---|----|---|
| 47 | 33 | 40 | 55 | 47 | 42 | 74 | 40 | 15 | 17 | 15:15 |

| 1 | 2 | 3 | 4 | 5 | 6 | 7 | 8 | 9 | 10 | |
|---|---|---|---|---|---|---|---|---|----|---|
| 43 | 59 | 86 | 31 | 64 | 30 | 49 | 69 | 43 | 49 | 15:16 |

WEEK 15 – SHAPE PUZZLES

= 4    = 8

= 11    = 6

# WEEK 16 – PRACTICE WORK

### DAY 1 – MONDAY

| 1 | 2 | 3 | 4 | 5 | 6 | 7 | 8 | |
|---|---|---|---|---|---|---|---|---|
| 55 | 66 | 77 | 88 | 99 | 79 | 46 | 88 | 16:1 |

| 1 | 2 | 3 | 4 | 5 | 6 | 7 | 8 | |
|---|---|---|---|---|---|---|---|---|
| 46 | 56 | 76 | 02 | 05 | 12 | 02 | 42 | 16:2 |

| 1 | 2 | 3 | 4 | 5 | 6 | 7 | 8 | |
|---|---|---|---|---|---|---|---|---|
| 41 | 31 | 00 | 03 | 98 | 04 | 83 | 46 | 16:3 |

| 1 | 2 | 3 | 4 | 5 | 6 | 7 | 8 | |
|---|---|---|---|---|---|---|---|---|
| 37 | 60 | 26 | 53 | 03 | 56 | 10 | 11 | 16:4 |

### DAY 2 – TUESDAY

| 1 | 2 | 3 | 4 | 5 | 6 | 7 | 8 | |
|---|---|---|---|---|---|---|---|---|
| 41 | 86 | 77 | 55 | 57 | 57 | 51 | 26 | 16:5 |

| 1 | 2 | 3 | 4 | 5 | 6 | 7 | 8 | |
|---|---|---|---|---|---|---|---|---|
| 03 | 28 | 14 | 75 | 00 | 00 | 58 | 14 | 16:6 |

### DAY 3 – WEDNESDAY

| 1 | 2 | 3 | 4 | 5 | 6 | 7 | 8 | |
|---|---|---|---|---|---|---|---|---|
| 58 | 14 | 23 | 88 | 32 | 87 | 75 | 77 | 16:7 |

| 1 | 2 | 3 | 4 | 5 | 6 | 7 | 8 | |
|---|---|---|---|---|---|---|---|---|
| 73 | 52 | 40 | 38 | 57 | 73 | 06 | 52 | 16:8 |

### DAY 4 – THURSDAY

| 1 | 2 | 3 | 4 | 5 | 6 | 7 | 8 | |
|---|---|---|---|---|---|---|---|---|
| 86 | 69 | 47 | 55 | 64 | 21 | 36 | 75 | 16:9 |

| 1 | 2 | 3 | 4 | 5 | 6 | 7 | 8 | |
|---|---|---|---|---|---|---|---|---|
| 74 | 25 | 02 | 99 | 36 | 33 | 74 | 88 | 16:10 |

### DAY 5 – FRIDAY

| 1 | 2 | 3 | 4 | 5 | 6 | 7 | 8 | |
|---|---|---|---|---|---|---|---|---|
| 84 | 31 | 36 | 89 | 64 | 68 | 23 | 80 | 16:11 |

| 1 | 2 | 3 | 4 | 5 | 6 | 7 | 8 | |
|---|---|---|---|---|---|---|---|---|
| 20 | 54 | 00 | 69 | 41 | 75 | 99 | 87 | 16:12 |

# WEEK 16 – MIND MATH PRACTICE WORK

### DAY 1 – MONDAY

| 1 | 2 | 3 | 4 | 5 | 6 | 7 | 8 | 9 | 10 | |
|---|---|---|---|---|---|---|---|---|---|---|
| 45 | 04 | 20 | 33 | 34 | 60 | 10 | 74 | 58 | 34 | 16:13 |

### DAY 2 – TUESDAY

| 1 | 2 | 3 | 4 | 5 | 6 | 7 | 8 | 9 | 10 | |
|---|---|---|---|---|---|---|---|---|---|---|
| 06 | 43 | 53 | 56 | 11 | 92 | 20 | 36 | 12 | 12 | 16:14 |

### DAY 3 – WEDNESDAY

| 1 | 2 | 3 | 4 | 5 | 6 | 7 | 8 | 9 | 10 | |
|---|---|---|---|---|---|---|---|---|---|---|
| 24 | 75 | 76 | 57 | 64 | 79 | 87 | 07 | 33 | 86 | 16:15 |

### DAY 4 – THURSDAY

| 1 | 2 | 3 | 4 | 5 | 6 | 7 | 8 | 9 | 10 | |
|---|---|---|---|---|---|---|---|---|---|---|
| 58 | 78 | 56 | 55 | 12 | 37 | 66 | 36 | 79 | 63 | 16:16 |

### DAY 5 – FRIDAY

| 1 | 2 | 3 | 4 | 5 | 6 | 7 | 8 | 9 | 10 | |
|---|---|---|---|---|---|---|---|---|---|---|
| 01 | 99 | 26 | 34 | 62 | 27 | 75 | 20 | 56 | 77 | 16:17 |

| 1 | 2 | 3 | 4 | 5 | 6 | 7 | 8 | 9 | 10 | |
|---|---|---|---|---|---|---|---|---|---|---|
| 38 | 77 | 17 | 85 | 54 | 00 | 57 | 52 | 28 | 33 | 16:18 |

WEEK 16 – SHAPE PUZZLES

= 5          = 7

= 8          = 3

## WEEK 17 – PRACTICE WORK
### DAY 1 – MONDAY

| 1 | 2 | 3 | 4 | 5 | 6 | 7 | 8 | |
|---|---|---|---|---|---|---|---|---|
| 17 | 02 | 22 | 15 | 23 | 14 | 44 | 33 | 17:1 |

| 1 | 2 | 3 | 4 | 5 | 6 | 7 | 8 | |
|---|---|---|---|---|---|---|---|---|
| 30 | 02 | 25 | 45 | 30 | 22 | 11 | 74 | 17:2 |

| 1 | 2 | 3 | 4 | 5 | 6 | 7 | 8 | |
|---|---|---|---|---|---|---|---|---|
| 20 | 10 | 61 | 31 | 95 | 24 | 33 | 30 | 17:3 |

| 1 | 2 | 3 | 4 | 5 | 6 | 7 | 8 | |
|---|---|---|---|---|---|---|---|---|
| 00 | 11 | 42 | 12 | 21 | 20 | 00 | 42 | 17:4 |

### DAY 2 – TUESDAY

| 1 | 2 | 3 | 4 | 5 | 6 | 7 | 8 | |
|---|---|---|---|---|---|---|---|---|
| 75 | 33 | 14 | 12 | 37 | 13 | 01 | 34 | 17:5 |

| 1 | 2 | 3 | 4 | 5 | 6 | 7 | 8 | |
|---|---|---|---|---|---|---|---|---|
| 04 | 49 | 01 | 45 | 31 | 81 | 23 | 02 | 17:6 |

### DAY 3 – WEDNESDAY

| 1 | 2 | 3 | 4 | 5 | 6 | 7 | 8 | |
|---|---|---|---|---|---|---|---|---|
| 06 | 86 | 10 | 02 | 15 | 74 | 50 | 46 | 17:7 |

| 1 | 2 | 3 | 4 | 5 | 6 | 7 | 8 | |
|---|---|---|---|---|---|---|---|---|
| 32 | 03 | 21 | 01 | 76 | 57 | 01 | 89 | 17:8 |

### DAY 4 – THURSDAY

| 1 | 2 | 3 | 4 | 5 | 6 | 7 | 8 | |
|---|---|---|---|---|---|---|---|---|
| 78 | 02 | 69 | 13 | 21 | 97 | 77 | 56 | 17:9 |

| 1 | 2 | 3 | 4 | 5 | 6 | 7 | 8 | |
|---|---|---|---|---|---|---|---|---|
| 41 | 13 | 02 | 59 | 34 | 88 | 34 | 31 | 17:10 |

### DAY 5 – FRIDAY

| 1 | 2 | 3 | 4 | 5 | 6 | 7 | 8 | |
|---|---|---|---|---|---|---|---|---|
| 21 | 24 | 95 | 67 | 13 | 00 | 04 | 21 | 17:11 |

| 1 | 2 | 3 | 4 | 5 | 6 | 7 | 8 | |
|---|---|---|---|---|---|---|---|---|
| 58 | 45 | 02 | 52 | 42 | 77 | 10 | 30 | 17:12 |

## WEEK 17 – MIND MATH PRACTICE WORK
### DAY 1 – MONDAY

| 1 | 2 | 3 | 4 | 5 | 6 | 7 | 8 | 9 | 10 | |
|---|---|---|---|---|---|---|---|---|----|---|
| 33 | 09 | 14 | 10 | 00 | 16 | 31 | 50 | 21 | 04 | 17:13 |

### DAY 2 – TUESDAY

| 1 | 2 | 3 | 4 | 5 | 6 | 7 | 8 | 9 | 10 | |
|---|---|---|---|---|---|---|---|---|----|---|
| 10 | 04 | 25 | 33 | 22 | 01 | 53 | 22 | 14 | 20 | 17:14 |

### DAY 3 – WEDNESDAY

| 1 | 2 | 3 | 4 | 5 | 6 | 7 | 8 | 9 | 10 | |
|---|---|---|---|---|---|---|---|---|----|---|
| 24 | 02 | 53 | 53 | 47 | 48 | 29 | 14 | 41 | 44 | 17:15 |

### DAY 4 – THURSDAY

| 1 | 2 | 3 | 4 | 5 | 6 | 7 | 8 | 9 | 10 | |
|---|---|---|---|---|---|---|---|---|----|---|
| 03 | 15 | 12 | 70 | 14 | 10 | 14 | 33 | 47 | 33 | 17:16 |

### DAY 5 – FRIDAY

| 1 | 2 | 3 | 4 | 5 | 6 | 7 | 8 | 9 | 10 | |
|---|---|---|---|---|---|---|---|---|----|---|
| 46 | 44 | 14 | 33 | 22 | 40 | 66 | 20 | 17 | 16 | 17:17 |

| 1 | 2 | 3 | 4 | 5 | 6 | 7 | 8 | 9 | 10 | |
|---|---|---|---|---|---|---|---|---|----|---|
| 22 | 44 | 43 | 11 | 21 | 23 | 30 | 41 | 23 | 40 | 17:18 |

### WEEK 17 – SHAPE PUZZLES

 = 6      = 8

= 11     = 12

www.abacus-math.com

## WEEK 18 – PRACTICE WORK

### DAY 1 – MONDAY

| 1 | 2 | 3 | 4 | 5 | 6 | 7 | 8 | |
|---|---|---|---|---|---|---|---|---|
| 35 | 23 | 02 | 75 | 43 | 24 | 23 | 10 | 18:1 |

| 1 | 2 | 3 | 4 | 5 | 6 | 7 | 8 | |
|---|---|---|---|---|---|---|---|---|
| 15 | 87 | 85 | 85 | 39 | 99 | 15 | 77 | 18:2 |

### DAY 2 – TUESDAY

| 1 | 2 | 3 | 4 | 5 | 6 | 7 | 8 | |
|---|---|---|---|---|---|---|---|---|
| 55 | 68 | 03 | 03 | 51 | 11 | 00 | 57 | 18:3 |

| 1 | 2 | 3 | 4 | 5 | 6 | 7 | 8 | |
|---|---|---|---|---|---|---|---|---|
| 12 | 96 | 89 | 68 | 98 | 10 | 84 | 22 | 18:4 |

### DAY 3 – WEDNESDAY

| 1 | 2 | 3 | 4 | 5 | 6 | 7 | 8 | |
|---|---|---|---|---|---|---|---|---|
| 16 | 71 | 77 | 95 | 66 | 41 | 78 | 33 | 18:5 |

| 1 | 2 | 3 | 4 | 5 | 6 | 7 | 8 | |
|---|---|---|---|---|---|---|---|---|
| 78 | 21 | 98 | 99 | 32 | 88 | 79 | 46 | 18:6 |

### DAY 4 – THURSDAY

| 1 | 2 | 3 | 4 | 5 | 6 | 7 | 8 | |
|---|---|---|---|---|---|---|---|---|
| 87 | 79 | 58 | 56 | 96 | 60 | 16 | 79 | 18:7 |

| 1 | 2 | 3 | 4 | 5 | 6 | 7 | 8 | |
|---|---|---|---|---|---|---|---|---|
| 43 | 79 | 12 | 59 | 99 | 41 | 10 | 41 | 18:8 |

### DAY 5 – FRIDAY

| 1 | 2 | 3 | 4 | 5 | 6 | 7 | 8 | |
|---|---|---|---|---|---|---|---|---|
| 04 | 55 | 77 | 74 | 16 | 20 | 15 | 74 | 18:9 |

| 1 | 2 | 3 | 4 | 5 | 6 | 7 | 8 | |
|---|---|---|---|---|---|---|---|---|
| 54 | 66 | 51 | 41 | 63 | 01 | 40 | 89 | 18:10 |

## WEEK 18 – MIND MATH PRACTICE WORK

### DAY 1 – MONDAY

| 1 | 2 | 3 | 4 | 5 | 6 | 7 | 8 | 9 | 10 | |
|---|---|---|---|---|---|---|---|---|---|---|
| 51 | 33 | 35 | 33 | 60 | 25 | 65 | 74 | 52 | 02 | 18:11 |

### DAY 2 – TUESDAY

| 1 | 2 | 3 | 4 | 5 | 6 | 7 | 8 | 9 | 10 | |
|---|---|---|---|---|---|---|---|---|---|---|
| 27 | 07 | 04 | 03 | 33 | 50 | 41 | 02 | 22 | 01 | 18:12 |

### DAY 3 – WEDNESDAY

| 1 | 2 | 3 | 4 | 5 | 6 | 7 | 8 | 9 | 10 | |
|---|---|---|---|---|---|---|---|---|---|---|
| 73 | 68 | 46 | 87 | 76 | 12 | 36 | 32 | 58 | 11 | 18:13 |

### DAY 4 – THURSDAY

| 1 | 2 | 3 | 4 | 5 | 6 | 7 | 8 | 9 | 10 | |
|---|---|---|---|---|---|---|---|---|---|---|
| 75 | 01 | 00 | 03 | 39 | 41 | 88 | 06 | 10 | 22 | 18:14 |

### DAY 5 – FRIDAY

| 1 | 2 | 3 | 4 | 5 | 6 | 7 | 8 | 9 | 10 | |
|---|---|---|---|---|---|---|---|---|---|---|
| 12 | 44 | 10 | 10 | 98 | 54 | 20 | 01 | 73 | 04 | 18:15 |

| 1 | 2 | 3 | 4 | 5 | 6 | 7 | 8 | 9 | 10 | |
|---|---|---|---|---|---|---|---|---|---|---|
| 36 | 54 | 62 | 56 | 56 | 10 | 02 | 04 | 00 | 10 | 18:16 |

WEEK 18 – SHAPE PUZZLES

= 10      = 6

= 7      = 9

## WEEK 19 – PRACTICE WORK
### DAY 1 – MONDAY

| 1 | 2 | 3 | 4 | 5 | 6 | 7 | 8 | |
|---|---|---|---|---|---|---|---|---|
| 06 | 20 | 31 | 00 | 50 | 03 | 414 | 446 | 19:1 |

| 1 | 2 | 3 | 4 | 5 | 6 | 7 | 8 | |
|---|---|---|---|---|---|---|---|---|
| 632 | 030 | 41 | 85 | 78 | 78 | 78 | 59 | 19:2 |

### DAY 2 – TUESDAY

| 1 | 2 | 3 | 4 | 5 | 6 | 7 | 8 | |
|---|---|---|---|---|---|---|---|---|
| 21 | 20 | 00 | 24 | 12 | 30 | 010 | 211 | 19:3 |

| 1 | 2 | 3 | 4 | 5 | 6 | 7 | 8 | |
|---|---|---|---|---|---|---|---|---|
| 462 | 120 | 52 | 78 | 14 | 72 | 75 | 62 | 19:4 |

### DAY 3 – WEDNESDAY

| 1 | 2 | 3 | 4 | 5 | 6 | 7 | 8 | |
|---|---|---|---|---|---|---|---|---|
| 32 | 15 | 14 | 02 | 77 | 30 | 474 | 203 | 19:5 |

| 1 | 2 | 3 | 4 | 5 | 6 | 7 | 8 | |
|---|---|---|---|---|---|---|---|---|
| 247 | 242 | 58 | 58 | 34 | 56 | 55 | 589 | 19:6 |

### DAY 4 – THURSDAY

| 1 | 2 | 3 | 4 | 5 | 6 | 7 | 8 | |
|---|---|---|---|---|---|---|---|---|
| 58 | 20 | 76 | 41 | 21 | 11 | 203 | 010 | 19:7 |

| 1 | 2 | 3 | 4 | 5 | 6 | 7 | 8 | |
|---|---|---|---|---|---|---|---|---|
| 576 | 021 | 55 | 86 | 56 | 40 | 765 | 213 | 19:8 |

### DAY 5 – FRIDAY

| 1 | 2 | 3 | 4 | 5 | 6 | 7 | 8 | |
|---|---|---|---|---|---|---|---|---|
| 00 | 66 | 00 | 04 | 10 | 25 | 53 | 19 | 19:9 |

| 1 | 2 | 3 | 4 | 5 | 6 | 7 | 8 | |
|---|---|---|---|---|---|---|---|---|
| 475 | 303 | 55 | 24 | 23 | 68 | 379 | 000 | 19:10 |

## WEEK 19 – MIND MATH PRACTICE WORK
### DAY 1 – MONDAY

| 1 | 2 | 3 | 4 | 5 | 6 | 7 | 8 | 9 | 10 | |
|---|---|---|---|---|---|---|---|---|---|---|
| 10 | 00 | 25 | 03 | 22 | 300 | 68 | 76 | 93 | 54 | 19:11 |

### DAY 2 – TUESDAY

| 1 | 2 | 3 | 4 | 5 | 6 | 7 | 8 | 9 | 10 | |
|---|---|---|---|---|---|---|---|---|---|---|
| 50 | 44 | 56 | 33 | 12 | 140 | 01 | 45 | 97 | 15 | 19:12 |

### DAY 3 – WEDNESDAY

| 1 | 2 | 3 | 4 | 5 | 6 | 7 | 8 | 9 | 10 | |
|---|---|---|---|---|---|---|---|---|---|---|
| 03 | 15 | 33 | 27 | 13 | 255 | 13 | 75 | 91 | 88 | 19:13 |

### DAY 4 – THURSDAY

| 1 | 2 | 3 | 4 | 5 | 6 | 7 | 8 | 9 | 10 | |
|---|---|---|---|---|---|---|---|---|---|---|
| 09 | 02 | 66 | 11 | 380 | 110 | 75 | 08 | 95 | 01 | 19:14 |

### DAY 5 – FRIDAY

| 1 | 2 | 3 | 4 | 5 | 6 | 7 | 8 | 9 | 10 | |
|---|---|---|---|---|---|---|---|---|---|---|
| 135 | 511 | 24 | 10 | 95 | 53 | 10 | 76 | 69 | 43 | 19:15 |

### WEEK 19 – SHAPE PUZZLES

 = 9    = 5

 = 3    = 0

## WEEK 20 – PRACTICE WORK
### DAY 1 – MONDAY

| 1 | 2 | 3 | 4 | 5 | 6 | 7 | 8 | 20:1 |
|---|---|---|---|---|---|---|---|---|
| 25 | 00 | 30 | 222 | 112 | 330 | 110 | 611 | |

| 1 | 2 | 3 | 4 | 5 | 6 | 7 | 8 | 20:2 |
|---|---|---|---|---|---|---|---|---|
| 410 | 312 | 30 | 79 | 947 | 828 | 27 | 000 | |

### DAY 2 – TUESDAY

| 1 | 2 | 3 | 4 | 5 | 6 | 7 | 8 | 20:3 |
|---|---|---|---|---|---|---|---|---|
| 96 | 64 | 32 | 762 | 513 | 63 | 558 | 712 | |

| 1 | 2 | 3 | 4 | 5 | 6 | 7 | 8 | 20:4 |
|---|---|---|---|---|---|---|---|---|
| 740 | 345 | 06 | 77 | 889 | 353 | 555 | 555 | |

### DAY 3 – WEDNESDAY

| 1 | 2 | 3 | 4 | 5 | 6 | 7 | 8 | 20:5 |
|---|---|---|---|---|---|---|---|---|
| 47 | 45 | 44 | 478 | 677 | 521 | 335 | 260 | |

| 1 | 2 | 3 | 4 | 5 | 6 | 7 | 8 | 20:6 |
|---|---|---|---|---|---|---|---|---|
| 133 | 254 | 69 | 102 | 343 | 485 | 40 | 221 | |

### DAY 4 – THURSDAY

| 1 | 2 | 3 | 4 | 5 | 6 | 7 | 8 | 20:7 |
|---|---|---|---|---|---|---|---|---|
| 45 | 89 | 42 | 63 | 31 | 57 | 34 | 86 | |

| 1 | 2 | 3 | 4 | 5 | 6 | 7 | 8 | 20:8 |
|---|---|---|---|---|---|---|---|---|
| 323 | 440 | 241 | 231 | 401 | 146 | 303 | 206 | |

### DAY 5 – FRIDAY

| 1 | 2 | 3 | 4 | 5 | 6 | 7 | 8 | 20:9 |
|---|---|---|---|---|---|---|---|---|
| 50 | 69 | 85 | 44 | 83 | 10 | 41 | 17 | |

| 1 | 2 | 3 | 4 | 5 | 6 | 7 | 8 | 20:10 |
|---|---|---|---|---|---|---|---|---|
| 141 | 76 | 240 | 31 | 233 | 11 | 786 | 544 | |

## WEEK 20 – MIND MATH PRACTICE WORK
### DAY 1 – MONDAY

| 1 | 2 | 3 | 4 | 5 | 6 | 7 | 8 | 9 | 10 | 20:11 |
|---|---|---|---|---|---|---|---|---|---|---|
| 25 | 21 | 50 | 158 | 24 | 78 | 38 | 55 | 85 | 37 | |

### DAY 2 – TUESDAY

| 1 | 2 | 3 | 4 | 5 | 6 | 7 | 8 | 9 | 10 | 20:12 |
|---|---|---|---|---|---|---|---|---|---|---|
| 57 | 52 | 24 | 710 | 200 | 36 | 93 | 69 | 59 | 67 | |

### DAY 3 – WEDNESDAY

| 1 | 2 | 3 | 4 | 5 | 6 | 7 | 8 | 9 | 10 | 20:13 |
|---|---|---|---|---|---|---|---|---|---|---|
| 87 | 77 | 58 | 54 | 043 | 92 | 74 | 86 | 32 | 40 | |

### DAY 4 – THURSDAY

| 1 | 2 | 3 | 4 | 5 | 6 | 7 | 8 | 9 | 10 | 20:14 |
|---|---|---|---|---|---|---|---|---|---|---|
| 135 | 111 | 26 | 400 | 33 | 69 | 43 | 12 | 45 | 02 | |

### DAY 5 – FRIDAY

| 1 | 2 | 3 | 4 | 5 | 6 | 7 | 8 | 9 | 10 | 20:15 |
|---|---|---|---|---|---|---|---|---|---|---|
| 52 | 110 | 40 | 110 | 98 | 63 | 91 | 54 | 76 | 49 | |

### WEEK 20 – SHAPE PUZZLES

= 12    = 7

= 8    = 10

## End of Level 1 Test

| 1 | 2 | 3 | 4 | 5 | 6 | 7 | 8 | 9 | 10 | |
|---|---|---|---|---|---|---|---|---|----|---|
| 57 | 77 | 98 | 79 | 97 | 98 | 27 | 69 | 46 | 88 | Test:1 |

| 1 | 2 | 3 | 4 | 5 | 6 | 7 | 8 | 9 | 10 | |
|---|---|---|---|---|---|---|---|---|----|---|
| 52 | 30 | 75 | 01 | 54 | 25 | 24 | 46 | 32 | 65 | Test:2 |

| 1 | 2 | 3 | 4 | 5 | 6 | 7 | 8 | 9 | 10 | |
|---|---|---|---|---|---|---|---|---|----|---|
| 75 | 45 | 42 | 75 | 21 | 54 | 71 | 41 | 21 | 20 | Test:3 |

| 1 | 2 | 3 | 4 | 5 | 6 | 7 | 8 | 9 | 10 | |
|---|---|---|---|---|---|---|---|---|----|---|
| 42 | 66 | 53 | 20 | 74 | 13 | 35 | 16 | 38 | 30 | Test:4 |

| 1 | 2 | 3 | 4 | 5 | 6 | 7 | 8 | 9 | 10 | |
|---|---|---|---|---|---|---|---|---|----|---|
| 22 | 02 | 61 | 13 | 01 | 12 | 22 | 41 | 34 | 33 | Test:5 |

| 1 | 2 | 3 | 4 | 5 | 6 | 7 | 8 | 9 | 10 | |
|---|---|---|---|---|---|---|---|---|----|---|
| 66 | 86 | 54 | 66 | 33 | 45 | 50 | 67 | 13 | 55 | Test:6 |

| 1 | 2 | 3 | 4 | 5 | 6 | 7 | 8 | 9 | 10 | |
|---|---|---|---|---|---|---|---|---|----|---|
| 45 | 63 | 87 | 43 | 41 | 04 | 77 | 22 | 76 | 66 | Test:7 |

| 1 | 2 | 3 | 4 | 5 | 6 | 7 | 8 | 9 | 10 | |
|---|---|---|---|---|---|---|---|---|----|---|
| 87 | 13 | 20 | 10 | 72 | 70 | 99 | 51 | 85 | 33 | Test:8 |

## MIND MATH

| 1 | 2 | 3 | 4 | 5 | 6 | 7 | 8 | 9 | 10 | |
|---|---|---|---|---|---|---|---|---|----|---|
| 76 | 11 | 50 | 53 | 99 | 76 | 63 | 47 | 46 | 37 | Test:9 |

| 1 | 2 | 3 | 4 | 5 | 6 | 7 | 8 | 9 | 10 | |
|---|---|---|---|---|---|---|---|---|----|---|
| 57 | 62 | 24 | 15 | 41 | 52 | 44 | 11 | 41 | 20 | Test:10 |

TEST – SHAPE PUZZLES

 = 4     ◎ = 10

 = 8     ☾ = 9

# ABOUT SAI SPEED MATH ACADEMY

One subject that is very important for success in this world, along with being able to read and write, is the knowledge of numbers. Math is one subject which requires proficiency from anyone who wants to achieve something in life. A strong foundation and a basic understanding of math is a must to mastering higher levels of math.

We, the family, best friends, and parents of children in elementary school, early on discovered that what our children were learning at school was not enough for them to master the basics of math. Teachers at school, with the resources they had, did the best they could. But, as parents, we had to do more to help them understand the relationship between numbers and basic functions of adding, subtracting, multiplying and dividing. Also, what made us cringe is the fact that our children's attitude towards more complex math was to say, "Oh, we are allowed to use a calculator in class". This did not sit well with us. Even though we did not have a specific system that we followed, each of us could do basic calculations in our minds without looking for a calculator. So, this made us want to do more for our children.

We started to look into the various methods that were available in the marketplace to help our children understand basic math and reduce their dependency on calculators. We came across soroban, a wonderful calculating tool from Japan. Soroban perfectly fits with the base-10 number system used at present and provides a systematic method to follow while calculating in one's mind.

This convinced us and within a short time we were able to work with fluency on the tool. The next step was to introduce it to our children, which we thought was going to be an easy task. It, however, was not. It was next to impossible to find the resources or the curriculum to help us introduce the tool in the correct order. Teaching all the concepts in one sitting and expecting children to apply them to the set of problems we gave them only made them push away the tool in frustration.

However, help comes to those who ask, and to those who are willing to work to achieve their goals. We came across a soroban teacher who helped us by giving us ideas and an outline of how soroban should be introduced. But, we still needed an actual worksheet to give our children to practice on. That is when we decided to come up with practice worksheets of our own design for our kids.

Slowly and steadily, practicing with the worksheets that we developed, our children started to get the idea and loved what they could do with a soroban. Soon we realized that they were better with mind math than we were.

Today, 6 years later, all our kids have completed their soroban training and are reaping the benefits of the hard work that they did over the years.

Now, although very happy, we were humbled at the number of requests we got from parents who wanted to know more about our curriculum. We had no way to share our new knowledge with them.

Now, through the introduction of our instruction book and workbooks, that has changed. We want to share everything we know with all the dedicated parents who are interested in teaching soroban to their children. This is our humble attempt to bring a systematic instruction manual and corresponding workbook to help introduce your children to soroban.

What started as a project to help our kids has grown over the years and we are fortunate to say that a number of children have benefitted learning with the same curriculum that we developed for our children.

Thank you for choosing our system to enhance your children's mathematical skills.

We love working on soroban and hope you do too!

# List of SAI Speed Math Academy Publications

## LEVEL – 1

Abacus Mind Math Instruction Book Level – 1: Step by Step Guide to Excel at Mind Math with Soroban, a Japanese Abacus
**ISBN-13:** 978-1941589007

Abacus Mind Math Level – 1 Workbook 1 of 2: Excel at Mind Math with Soroban, a Japanese Abacus

**ISBN-13:** 978-1941589014

Abacus Mind Math Level – 1 Workbook 2 of 2: Excel at Mind Math with Soroban, a Japanese Abacus

**ISBN-13:** 978-1941589021

## LEVEL – 2

Abacus Mind Math Instruction Book Level – 2: Step by Step Guide to Excel at Mind Math with Soroban, a Japanese Abacus
**ISBN-13:** 978-1941589038

Abacus Mind Math Level – 2 Workbook 1 of 2: Excel at Mind Math with Soroban, a Japanese Abacus

**ISBN-13:** 978-1941589045

Abacus Mind Math Level – 2 Workbook 2 of 2: Excel at Mind Math with Soroban, a Japanese Abacus

**ISBN-13:** 978-1941589052

## LEVEL – 3

Abacus Mind Math Instruction Book Level – 3: Step by Step Guide to Excel at Mind Math with Soroban, a Japanese Abacus
**ISBN-13:** 9781941589069

Abacus Mind Math Level – 3 Workbook 1 of 2: Excel at Mind Math with Soroban, a Japanese Abacus

**ISBN-13:** 9781941589076

Abacus Mind Math Level – 3 Workbook 2 of 2: Excel at Mind Math with Soroban, a Japanese Abacus

**ISBN-13:** 9781941589083

Continued.....

## <u>*Not Related To The Abacus Mind Math Series Books*</u>

**Learn To Do MATH WITH
SOROBAN, a Japanese Abacus**

**ISBN-13:** 978-1-5371-6329-1